USES OF ASME BOILER & PRESSURE VESSEL CODES

General Overview of
PRESSURE VESSEL

Writer
Chetan Singh
QA/QC Engineer

General Overview of PRESSURE VESSEL

WHAT IS A PRESSURE VESSEL ?
- Pressure Vessel is a container for the containment of pressure, either internal or external.

This pressure may be obtained from an external source, or by The application of heat from a direct or indirect source, or any combination thereof.

- Classification based on
  - Function
    - Storage Tank, Process vessel, Heat Exchanger
  - Geometry
    - Cylindrical, Spherical, Conical, Non-circular, Horizontal/vertical
  - Construction
    - Monowall, Multiwall, Forged, Cast
  - Service
    - Cryogenic, Steam, Lethal, Vacuum, Fired/Unfired, Stationary/Mobile

PARTS OF PRESSURE VESSEL
- DISH END (Blank diameter, Petal size, Nominal thickness)
  - Types (Spherical, Conical, Elliptical, Torispherical)
- SHELL (TL to TL, WL to WL, Straight Face)
  - Cladding of shell and nozzles
- Nozzle (forged, fabricated)
- Tray support rings, bolting bars, down- comer
- Nozzle reinforcement pads
- External attachments (ladder, pipe cleats)
- Lugs (for earthing, lifting, tailing)
- Saddle support / skirt support
- Pipe or manway davit
- Ladder rungs
- Internals (demister, vortex breaker, deflector etc)
- Name plate & asme stamps

SUPPORTS FOR VESSEL
- Supports are required for the installation of the vessel on the foundation.
- Type of supports
  - Leg type support
  - Skirt type support (cylindrical/conical) – using a d markblatter
  - Bracket type support
  - Saddle support – this support design using zick analysis

DESIGN CONSIDERATIONS
Following Loadings to be Considered Loadings to be considered in Designing the Vessels as per ASME SEC. VIII DIV. I (CLAUSE UG - 22):
(a) Internal or External Design pressure
(b) Weight of the vessel and normal contents under operating or test conditions (This includes additional Pressure due to static head of liquids);
(c) Superimposed static reactions from weight of attached equipment, such as motors, machinery, other vessels, piping, linings, and insulation;
(d) The attachment of:
    (1) internals (see Appendix D);
    (2) vessel supports, such as lugs, rings, skirts, saddles, and legs (see Appendix G);
(e) Cyclic and dynamic reactions due to pressure or thermal variations, or from equipment mounted on a vessel, and mechanical loadings;
(f) Wind, snow, and seismic reactions, where required;
(g) Impact reactions such as those due to fluid shock;
(h) Temperature gradients and differential thermal expansion.

G.A. DRAWING- ELEVATION

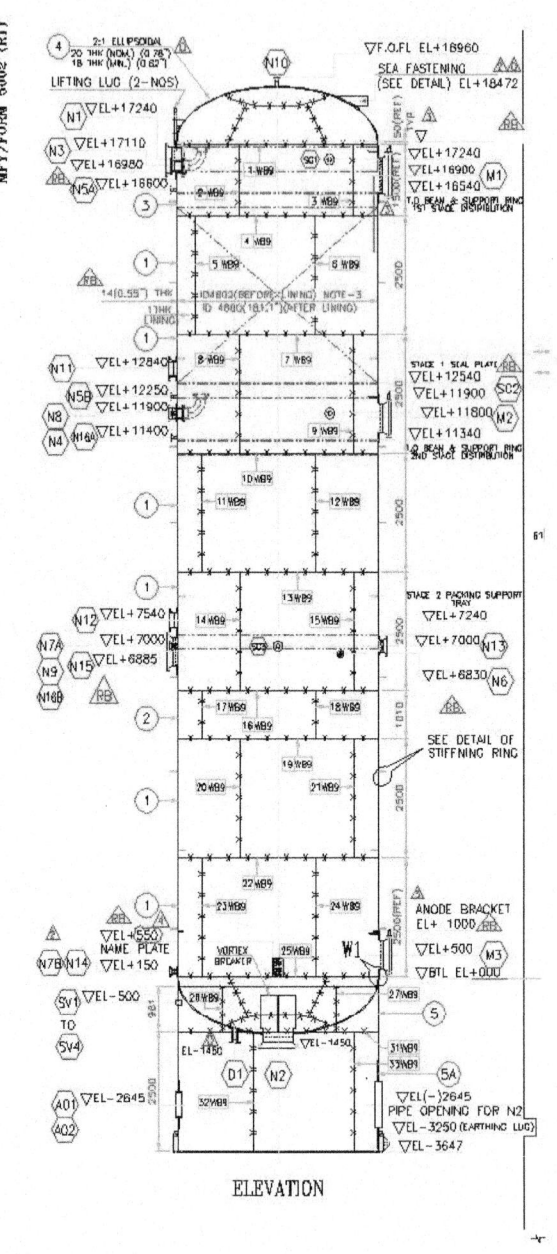

ELEVATION

General Arrangement Drawing

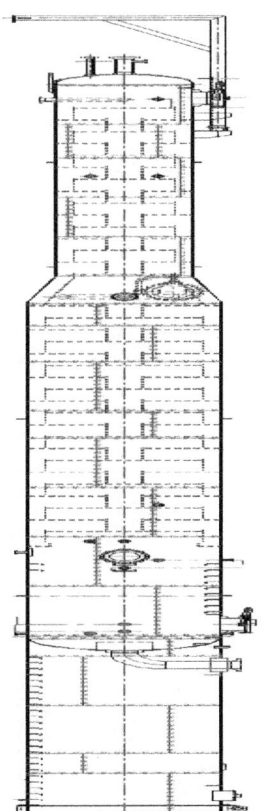

PLAN

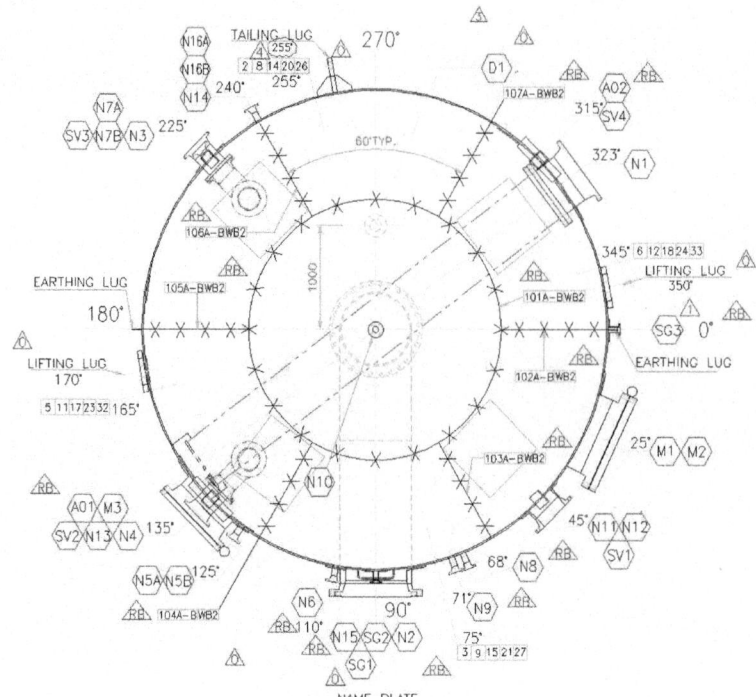

NOZZLE ORIENTATION PLAN

SKIRT DETAIL

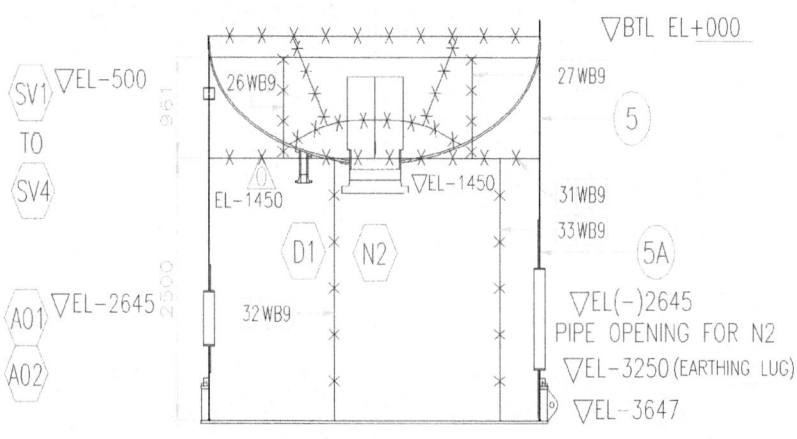

ELEVATION

Detail of Skirt

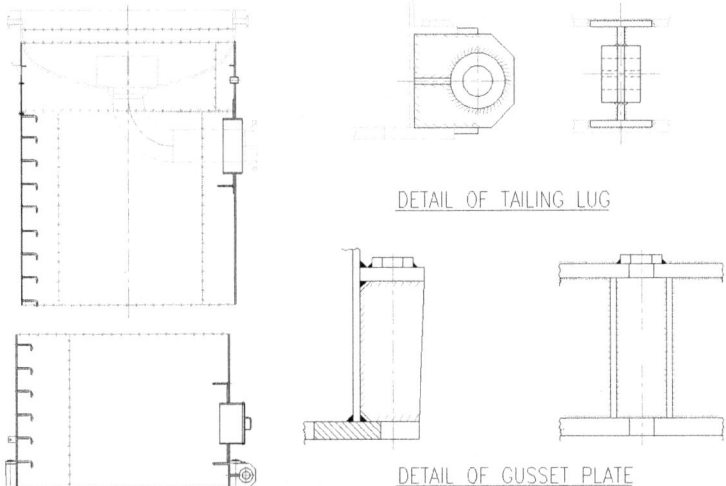

DETAIL OF TAILING LUG

DETAIL OF GUSSET PLATE

HEADS / END CLOSURES

Head is the part which closes the end opening of the cylindrical shell.

Types of the heads:
- Bolted blind flange
- Conical head
- Elliptical head
- Hemispherical head
- Toriconical head
- Torispherical head

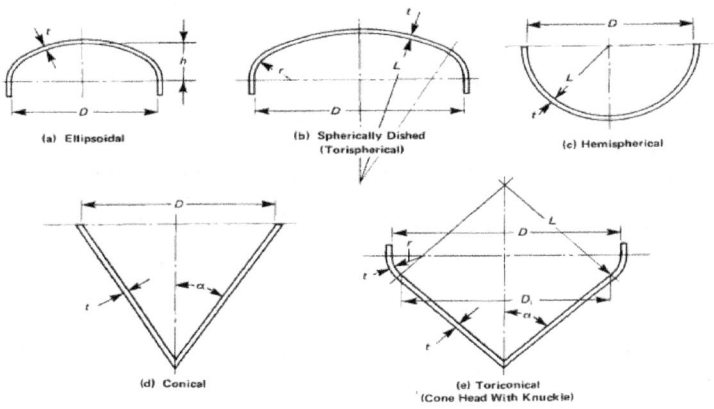

(a) Ellipsoidal
(b) Spherically Dished (Torispherical)
(c) Hemispherical
(d) Conical
(e) Toriconical (Cone Head With Knuckle)

FIG. 1-4 PRINCIPAL DIMENSIONS OF TYPICAL HEADS

NOZZLES / CONNECTIONS

- Nozzles are the openings provided in the shell / head for connecting the external piping with vessel.
- Nozzles consists of nozzle neck, flange & rf pad
- Types of the nozzles / connections :
- Pipe neck with flange connection
- Forged neck with flange connection
- Stub end connection
- Nozzle neck to be designed as per asme sec viii div i cl. Ug-45
- Rf pad for the nozzle to be designed as per asme sec viii div i cl. Ug-36
- Non std. Flange to be designed as per asme sec viii div i, appx-2 & std flange to be selected as per ug-44/asme b16.5

Nozzle Details

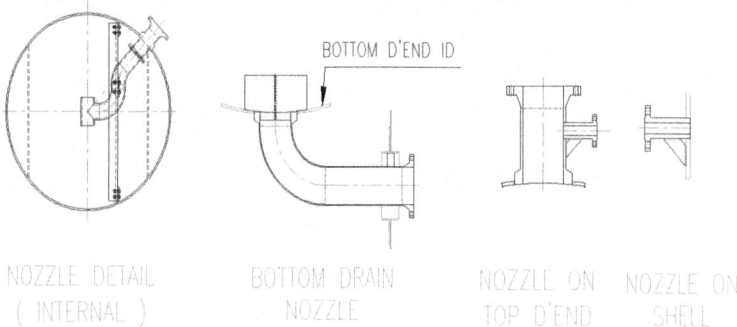

NOZZLE DETAIL (INTERNAL) BOTTOM DRAIN NOZZLE NOZZLE ON TOP D'END NOZZLE ON SHELL

Detail of Manhole

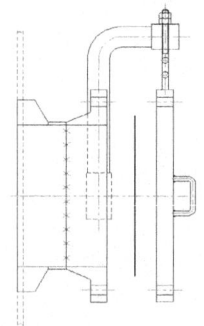 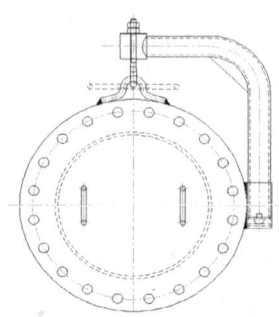 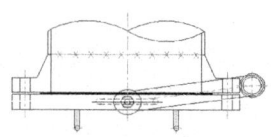

TSR Drawings

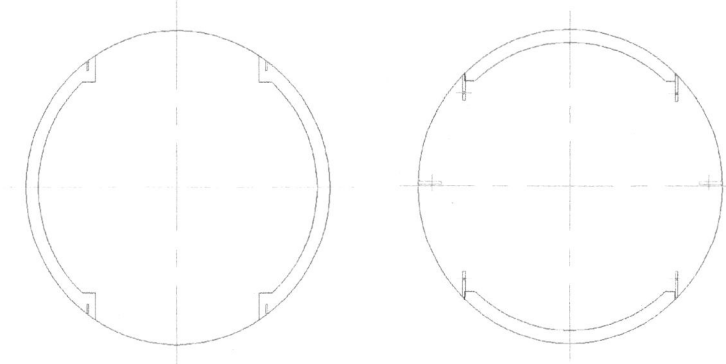

Support Cleats

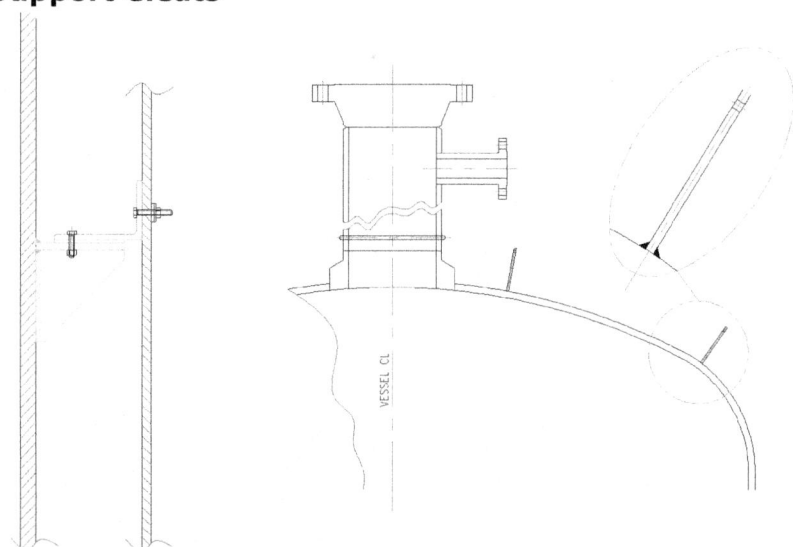

Name Plate Detail

	Company Name	
DESIGNED BY		company name
MANUFACTURED FOR		
ITEM NUMBER		
MANUFACTURER SERIAL NO.		
YEAR OF FABRICATION	2005	
CODES	ASME SEC. VIII DIV.1 ED.2001. ADD.02	INSPECTED BY
DESIGN PRESSURE	INT:15.82 KG/CM²(g), EXT:[V]	TEST PRESSURE (HORIZONTAL) 23.77 Kg/CM²(g)
DESIGN TEMPERATURE	INT:232°C, EXT.171°C	DATE OF TEST
CAPACITY	144.5M³	CORROSION ALLOW. 3 MM
OPERATING FLUID	HYDROCARBON	RADIOGRAPHY 100%
TOTAL WEIGHT EMPTY	64200 Kg	HEAT TREATMENT YES

Shell Development

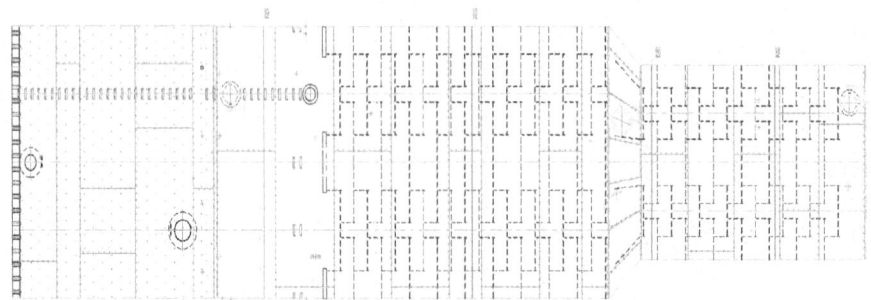

EQUIPMENT DESIGN IN SOFTWARE

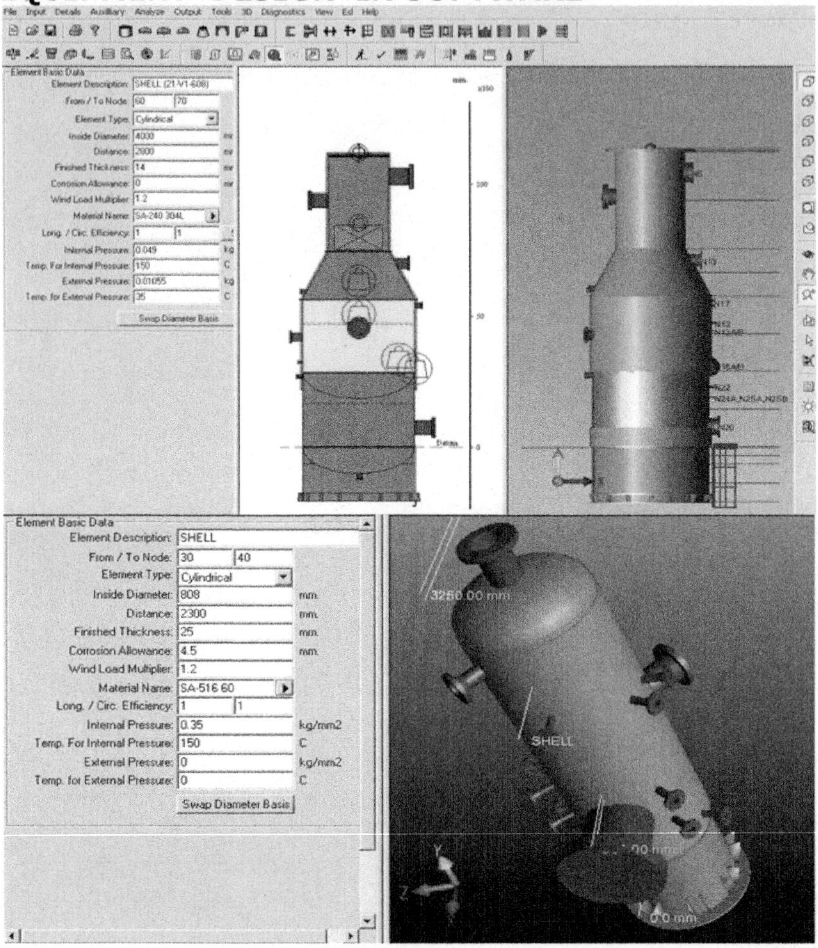

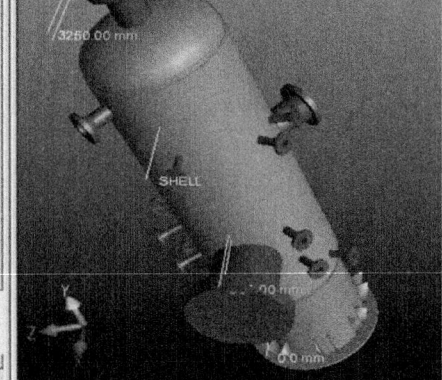

Material Selection

The following table gives general guidelines for material selection for various Pressure Parts / Non Pressure Parts of the equipment:-

DESIGN TEMP.°C	PRESSURE PARTS					NON PRESSURE PARTS			
	PLATE	PIPE	FORGING	BOLTS/ STUDS/ NUTS EXTERNAL	STRUCTURAL ATTACHMENT WELDED TO PRESSURE PARTS	INTERNAL PIPES	STUDS/ BOLTS/ NUTS INTERNAL
LOW TEMPERATURE							
ABOVE -80 UP TO -60	SA 203 GR.B IMPACT TESTED	SA 333 GR.3	SA 350 GR.LF3	SA 320 L7 SA 194 GR.4 OR GR.7	SA 203 GR.B	SA 333 GR.3	SA 193 GR.B8 SA 194 GR.8
ABOVE -60 UP TO -45	SA 537 CL.1 IMPACT TESTED	SA 333 GR.3	SA 350 GR.LF3	SA 320 L7 SA 194 GR.4 OR GR.7	SA 537 CL.1	SA 333 GR.3	SA 193 GR.B8 SA 194 GR.8
ABOVE -45 UP TO -29	SA 516 (ALL GRADES) IMPACT TESTED	SA 333 GR.6 OR GR.1	SA 350 GR.LF2	SA 320 L7 SA 194 GR.4 OR GR.7	SA 516 (IN ALL GRADES)	SA 333 GR.6	SA 193 GR.B8 SA 194 GR.8
ABOVE -29 UP TO 0	SA 516 (ALL GRADES)	SA 106 GR.B	SA 105/ SA 266	SA 193 GR.B7 SA 194 GR.2H	SA 516 (IN ALL GRADES)	SA 106 GR.B	SA 193 GR.B8 SA 194 GR.8
INTERMIDIATE TEMPERATURE							
ABOVE 0 UP TO 343	SA 516 (ALL GRADES)	SA 106 GR.B SA 312 TP 304L 316L,321 SA-376 TP 321	SA 105 SA 266	SA 193 B7 SA 194 GR.2H	IS 2062 (PLATES)	SA 106 GR.B	SA 193 GR.B8 SA 194 GR.4
	SA 240 TYPE 304L, 316L, 321		SA 182 F 304L,316L, 321	SA 193 B7 SA 194 GR.2H	SAME AS PRESSURE PARTS	SA 106 GR.B	SA 193 GR.B8 SA 194 GR.8
ABOVE 343 UP TO 427	SA 204 GR.B	SA 335 GR. P1	SA 182 GR.F1	SA 193 GR.B7 SA 194 GR.4	SAME AS PRESSURE PARTS	SAME AS PRESSURE PARTS	SA 193 GR.B8 SA 194 GR.4
	SA 387 GR.11 CL.1/CL.2	SA 335 GR.P11	SA 182 GR.F11	SA 193 GR.B7 SA 194 GR.4	SAME AS PRESSURE PARTS	SAME AS PRESSURE PARTS	SA 193 GR.B8 SA 194 GR.8
	SA 240 TYPE 304L,316L, 321	SA 312 TYPE 304L, 306L, 321,SA 276 TYPE 321	SA 182 F 304L, 306L, 321	SA 193 GR.B7 SA 194 GR.4	SAME AS PRESSURE PARTS	SAME AS PRESSURE PARTS	SA 193 GR.B8 SA 194 GR.8

SOUR SERVICE REQIREMENTS:

This shall be in accordance with NACE Standard MR-01-75 and shall meet the following special requirements for CS Material.

- All steel shall be fully killed and fine grained.
- All steel shall be produced using Basic oxygen or Electro Furnace Process.
- Steel shall be made by low sulphur and low phosphorous refining process and shall be vacuum degassed while molten by means of an approved procedure.
- Effective sulfide shape control by calcium treatment, if the sulphur level is in excess of 0.002%.

HYDROGEN INDUCED CRACKING (HIC) TESTING:

This shall be carried out on finished product of each heat irrespective of size/thickness as per NACE-TM-02-84 standards.

- The test shall be performed on a set of three test specimens. The test shall be performed, as per NACE-TM-02-84 and the acceptance criteria shall be as Crack sensitivity ratio (CSR) < 0.009% & Crack Length Ratio (CLR) < 10.00%
- In case any one of the above samples fails to meet the acceptance criteria, three more additional specimen from the product from which the first set of specimen were taken, shall be retested and results reported.
- In case of failure of any of the samples in above, two additional products shall be selected from the same heat and size specimens shall be tested (three from each product).
- In case of failure of any one of the six-samples, the particular heat will be rejected.

SULPHIDE STRESS CORROSION CRACKING (SSCC) TEST:

- This test shall be carried out on one finished product as well as on raw material of each heat (material wise and type of construction wise i.e. seamless and welded separately) irrespective of size/thickness. The test shall be carried out as per NACE-TM-01-77
- Reporting of test result: Curve shall be reported as per NACE-TM-01-77 for various stress level between 72% and 90% of SMYS.
- Acceptance Criteria: At 72% SMYS, time of failure shall not be less than 720 hrs.

INTERGRANULAR CORROSION (IGC) TEST:

The test shall be carried out for all Stainless Steel material, as per ASTM A 262 Practice E. The bend test specimen shall be examined at a magnification of 200x. This test shall be carried out on finished product as well as on each heat irrespective of size & thickness. The entire test reports shall be from the product specimen drawn from the heat bearing same number.

TESTING
- ALL BUTT JOINTS TO BE RT / UT / MPT / DP TESTED AS PER ASME CODE REQUIREMENT
- ALL CORNER JOINTS e.g. NOZZLE TO SHELL JOINT SHALL BE MPT / DP TESTED
- ALL FILLET WELDS SHALL BE DP TESTED
- ACCEPTANCE CRITERIA FOR NDT REQUIREMENT FOR WELD JOINT
  - FOR R.T ----- APPENDIX IV - ASME SEC. VIII DIV.1
  - FOR M.P.T ------ APPENDIX VI
  - FOR D.P.T ------ APPENDIX VIII
  - FOR U.T --------- APPENDIX XII
- EQUIPMENT TEST :
  - HYDROTEST -- 1.3 x DESIGN PRESSURE (UG - 99)
  - PNEUMATIC TEST --- 1.1 x DESIGN PRESSURE (UG - 100)
- OFFICIAL STAMP FOR CODED VESSELS
  - 'U' STAMPS --- ASME SEC. VIII DIV.1
  - 'UV' STAMPS --- FOR PRESSURE RELIEF VALVE

ASME BOILER & PRESSURE VESSEL CERTIFICATES OF AUTHORIZATION & CODE SYMBOL STAMPS

Sr. No.	SYMBOL/STAMP	CODE BOOKS REQUIRED
1	"A" (ASSEMBLY BOILER)	B 31.1 &SEC- 1 & SEC II –C
2	"U"(PRESSURE VESSEL)	ASME SEC VIII DIV - 1
3	"U2"(PRESSURE VESSEL)	ASME SEC VIII DIV-2
4	"U3" (HIGH PV)	ASME SEC VIII DIV-3
5	"S" (POWER BOILER)	B31.1 & SEC-1
6	"UV" (PV SAFETY VALVE)	ASME SEC VIIII DIV -1 /2
7	"V" (BOILER SAFETY VALVE)	SEC -1
8	"qp" (PRESSURE PIPING)	B31.1 & SEC -1

ASME Section VIII Code Stamps
"U"   (PRESSURE VESSEL) ---- ASME SEC VIII DIV -1
"U2" (PRESSURE VESSEL) ---- ASME SEC VIII DIV-2
"U3" (HIGH PV)   ----- ASME SEC VIII DIV- 3
"S"   (POWER BOILER) ------ B31.1 & SEC-1
"UV" (PV SAFETY VALVE)----ASME SEC VIIII DIV -1 /2
"V"   (BOILER SAFETY VALVE) ----- SEC –1
"R"   Repair Stamp

ASME BOILER & PRESSURE VESSELS CODES

- I. POWER BOILERS.
- II. MATERIAL SPECIFICATIONS.
- III. NUCLEAR POWER PLANT COMPONENTS.
- IV. HEATING BOILERS.
- V. NONDESTRUCTIVE EXAMINATION
- VI. RECOMMANDATED RULES FOR CARE & OPERATION OF HEATING BOILERS.
- VII. RECOMMANDATED RULE FOR CARE OF POWER BOILERS.
- VIII. PRESSURE VESSELS –DIVISION-1, DIVISION-2 & DIVISION-3.
- IX. WELDING & BRAZING QUALIFICATIONS.
- X. FIBERGLASS-REINFORCED PLASTIC PRESSURE VESSELS.
- XI. RULES FOR INSERVICE INSPECTION OF NUCLEAR POWER PLANT COMPONENTS.
- XII. RULES FOR CONSTRUCTION AND CONTINUED SERVICE OF TRANSPORT TANKS.

A Brief Discussion on ASME Section VIII Divisions 1 and 2 and Division 3.

	Section VIII Div.1	Section VIII Div.2	Section VIII Div.3
	"Unfired" Pressure Vessel Rules	Alternative Rules	Alternative Rules for High Pressure
Published	<1940	1968	1997
Pressure Limits	Normally up to 3000 psig	No limits either way, usually 600+ psig	No limits, Normally from 10,000 psig
Organisation	General, Construction Type & Material U, UG, UW, UF, UB, UCS, UNF, UCI, UCD, UHT, ULT	General, Material, Design, Fabrication and others AG, AM, AD, AF, AR, AI, AT, AS	Similar to Division 2 KG, KM, KD, KF, KR, KE, KT, KS
Design Factor	Design Factor 3.5 on tensile and other yield and temperature considerations	Design Factor 3 on tensile (lower factor under reviewed) and other yield and temperature considerations	Yield based with reduction factor for yield to tensile ratio less than 0.7

	Section VIII Div.1	Section VIII Div.2	Section VIII Div.3
	"Unfired" Pressure Vessel Rules	Alternative Rules	Alternative Rules for High Pressure
Design Rules	Membrane-Maximum stress Generally Elastic analysis Very detailed design rules with Quality (joint efficiency) Factors. Little stress analysis required, pure membrane without consideration of discontinuities controlling stress concentration to a safety factor of 3.5 or higher	Shell of Revolution – Max. Shear stress Generally Elastic Analysis Membrane + Bending. Fairly detailed design rules. In addition to the design rules, discontinuities, fatigue and other stress analysis considerations may be required unless exempted and guidance provided for in Appendix 4, 5 and 6	Maximum shear stress Elastic/plastic Analysis and more. Some design rules provided; Fatigue analysis required, Fracture mechanics evaluation required unless proven leak-before-burst, Residual stresses become significant and may be positive factors
Experimental stress Analysis	Normally not required	Introduced and may be required	Experimental design verification but may be exempted

	Section VIII Div.1	Section VIII Div.2	Section VIII Div.3
	"Unfired" Pressure Vessel Rules	Alternative Rules	Alternative Rules for High Pressure
Material and Impact Testing	Few restriction on materials; Impact required unless exempted, extensive exemptions under UG-20, UCS 66/67	More restriction of materials; impact required in general with similar rules as Division 1	Even more restrictive than Division 2 with different requirements. Fracture toughness testing requirement for fracture mechanics evaluation crack tip opening displacement (CTOD) testing and establishment of Klc and/or Jlc values
NDE Requirements	NDE requirements may be exempted through increased design factor	More stringent NDE requirements, extensive use of RT, as well as UT, MT and PT.	Even more restrictive than Division 2, UT used for all butt welds, RT otherwise, extensive use of PT and MT

	Section VIII Div.1	Section VIII Div.2	Section VIII Div.3
	"Unfired" Pressure Vessel Rules	Alternative Rules	Alternative Rules for High Pressure
Welding and Fabrication	Different types with butt welds and others	Extensive use/requirement of butt welds and full penetration welds including non pressure attachment welds	Butt welds and extensive use of other construction methods such as threaded, layered, wire-wound, interlocking strip wound and others
User	User or designated agent to provide specifications (see U-2(a))	User's Design Specification with detailed design requirements (see AG-301.1) include AD 160 for fatigue evaluation	User's Design specification with more specific details (see KG-310) including contained fluid data, etc. with useful operation life expected and others. Designer defined.

	Section VIII Div.1	Section VIII Div.2	Section VIII Div.3
	"Unfired" Pressure Vessel Rules	Alternative Rules	Alternative Rules for High Pressure
Manufacturer	Manufacturer to declare compliance in data report	Manufacturer's Design Report certifying design specification and code compliance in addition to data report	Same as Division 2
Professional Engineer Certification	Normally not required	Professional Engineers' Certification of User's Design Specification as well as Manufacturer's Design report Professional Engineer shall be experienced in pressure vessel design	Same as Division 2 but the Professional Engineer shall be in high-pressure vessel design and shall not sign for both User and Manufacturer.

	Section VIII Div.1	Section VIII Div.2	Section VIII Div.3
	"Unfired" Pressure Vessel Rules	Alternative Rules	Alternative Rules for High Pressure
Safety Relief Valve	UV Stamp	UV Stamp	UV3 Stamp
Code Stamp and Marking	U Stamp with Addition marking including W, P, RES, L, UB, DF, RT, HT	U2 Stamp with Additional marking including HT	U3 Stamp with additional marking denoting construction type, HT, PS, WL, M, F, W, UQT, WW, SW
Hydrostatic Test	1.3 (was 1.5 before the use of the 3.5 Design factor in the 1999 Addenda)	1.25	1.25

World Wide Pressure Vessel Codes

- Different industrial nations, institutions and organizations have developed standards and codes of the pressure vessel for
  - Design
  - Fabrication
  - Inspection

These codes and standards help the design engineer to size the vessel properly for its safe operation.

- Advantages of design codes:
  - Proven design based on experience.
  - Inbuilt factor of safety.
  - Amendments of codes are being done at regular interval based on feedback data, design improvements and technological up gradation in materials and fabrication processes.

- Country US:
  - Code: ASME Boiler and Pressure Vessel Code
    - ASME Section I: Power Boilers
    - ASME Section III: Nuclear Power Plant Components
    - ASME Section VIII Division-1, 2 & 3: Pressure Vessels
  - Issuing authority: American Society of Mechanical Engineers
- Country: UK
  - Code:
    - PD 5500: Unfired Fusion Welded Pressure Vessels
  - Issuing authority: British standard Institute
- Country: Germany
  - Code:
    - AD Merkblatter Technical Rules for Pressure Vessels
  - Issuing authority: Arbeitsgemeinschaft Druckbehalter

- Country: India
  - Code:
    - IS 2825: Code for Unfired Pressure Vessels
  - Issuing authority: Bureau of Indian Standards
- Country: Australia
  - Code:
    - AS 1200: SAA Boiler Code
    - AS 1210: Unfired Pressure Vessels
  - Issuing authority: Standards Association of Australia

- Country: USA
  - Code:
    - TEMA: Tubular Exchanger Manufacturers Association
  - Issuing authority: TEMA

IS 2825: Code for Unfired Pressure Vessels
- Issued in 1969 by Indian Standards Institution
- Code Scope
  - Pressure should be 1 kgf/cm2 < P > 200 kgf/cm2
  - D_o/D_i < 1.5
  - I.D > 150 mm
  - Not for Steam boiler, Steam and feed pipes
  - Not for Vessel in which Pr is due to static head only
  - Not for nuclear energy application
  - Hot water supply storage tanks if exceed
    - Heat input of 50000 kcal/h
    - Water capacity of 500 liters
    - Water temp of 110 °C
- Pressure vessels are classified in three classes:
  - Class 1:
    - Vessels which handles lethal and toxic substances.
    - Vessels in low temperature applications (-20° C and below).
    - Full radiography is required. (Joint efficiency = 1).

- Class 2:
  - Vessels which are not covered in class 1 or class 3.
  - Spot radiography is required (Joint efficiency = 0.85)
  - Maximum thickness limit = 38 mm including corrosion allowance.

- Class 3:
  - Light duty pressure vessels having shell thickness 16 mm or less.
  - Design pressure:
    - 3.5 kg/cm² or less for gaseous/vapor service.
    - 17.5 kg/cm² or less for liquid handling service
  - Design temperature: 0 °C to 250 °C
  - Radiography is not required (Joint efficiency 0.7)

- Allowable Design Stress:
  - Allowable design stress is minimum of following:
    - = (yield stress or 1% proof stress at design temperature for high alloy steel)/1.5
    - =(UTS at room temperature)/3
    - =(UTS at room temperature for high alloy steel)/2.5
    - =(Average rupture stress for rupture in 10^5 hrs)1.5
    - =(Average stress to produce 1% creep in 10^5 hrs)/1.0
  - Allowable stress values are given in table A.1, to A.4

PD 5500: Unfired Fusion Welded Pressure Vessels

- Issued in 1976 by British Standard Institute
- Organization of Code
  - Section 1 : General
  - Section 2 : Material
  - Section 3 : Design
  - Section 4 : Manufacture & Workmanship
  - Section 5 : Inspection & Testing
  - Annexes
  - Supplements
- Pressure vessels are classified in three categories:
  - Category 1: Vessels of all thicknesses with 100% radiography
  - Category 2: Vessels with spot radiography
    - Carbon & C-Mn steels up to 40 mm thickness
    - Alloy steel up to 30 mm thickness
    - Austenitic stainless steel and aluminum steel up to 40 mm thickness.
  - Category 3: Vessels subjected to no radiography but only visual inspection
    - Carbon & C-Mn steel up to 16 mm thickness
    - Austenitic stainless steel up to 25 mm thickness.
    - Upper temperature limit is 300 °C
    - Lower temperature limit is 0 °C for C & C-Mn steel.
- Design strength values for category 1 & 2:
  - Design strength value for Carbon, C-Mn and Low alloy steel are calculated as follow: (For material with specified elevated temperature values)
    - For temperature up to 50 °C: Minimum of following values:
      - =(Yield strength at room temperature)/1.5
      - =(UTS at room temperature)/2.35
      - =(Rupture stress for time t at design temperature)/1.3

- For temperature above 150 °C: Minimum of following values:
    - =(Yield strength at room temperature)/1.5
    - =(UTS at room temperature)/2.35
    - =(Rupture stress for time t at design temperature)/1.3
- Design strength values for category 1 & 2:
    - Design strength value for Carbon, C-Mn and Low alloy steel are calculated as follow: (For material without specified elevated temperature values)
        - For temperature up to 50 °C: Minimum of following values:
            - =(Yield strength at room temperature)/1.5
            - =(UTS at room temperature)/2.35
            - =(Rupture stress for time t at design temperature)/1.3
        - For temperature above 150 °C: Minimum of following values:
            - =(Yield strength at room temperature)/1.6
            - =(UTS at room temperature)/2.35
            - =(Rupture stress for time t at design temperature)/1.3
- Design strength values for category 1 & 2:
    - Design strength value for Austenitic Stainless Steel are calculated as follow: (For material with specified elevated temperature values)
        - For temperature up to 50 °C: Minimum of following values:
            - =(Yield strength at room temperature)/1.5
            - =(UTS at room temperature)/2.5
            - =(Rupture stress for time t at design temperature)/1.3

- For temperature above 150 °C: Minimum of following values:
  - =(Yield strength at room temperature)/1.5
  - =(UTS at room temperature)/2.5
  - =(Rupture stress for time t at design temperature)/1.3
- Design strength values for category 1 & 2:
  - Design strength value for Austenitic Stainless Steel are calculated as follow: (For material without specified elevated temperature values)
    - For temperature up to 50 °C: Minimum of following values:
      - =(Yield strength at room temperature)/1.5
      - =(UTS at room temperature)/2.5
      - =(Rupture stress for time t at design temperature)/1.3
    - For temperature above 150 °C: Minimum of following values:
      - =(Yield strength at room temperature)/1.45
      - =(UTS at room temperature)/2.5
      - =(Rupture stress for time t at design temperature)/1.3
- Design strength values for category 1 & 2:
  - Values between 50 °C and 150 °C shall be calculated by linear interpolation between values as calculated above.
- Design strength values for category 1 & 2:
  - For carbon and C-Mn Steel:
    - UTS at room temperature/5
  - For austenitic Stainless Steel:
    - Minimum of 120 N/mm^2 or 120*(450/(400+design temp °C))

- If minimum specified yield strength is less than 230 N/mm²
    - 0.8 x design strength calculated as per formulas for category 1 & 2.

- Design strength value tables:
    - Design strengths of different sets of pressure vessel materials at different temperatures are given in tables 2.3(a) to 2.3(k) and table 2A.3
    -

AD Merkblatter: Technical Rules for Pressure Vessels
- AD Merkblatter are prepared by the seven associations who together form AD.
- Organization of Code
    - Series A: Equipment installation and marking
    - Series B: Design
    - Series G: Fundamentals
    - Series H: Manufacture
    - Series HP: Manufacture and Testing
    - Series N: Non-metallic materials
    - Series S: Special cases
    - Series W: Metallic materials
- Pressure vessels are classified in seven groups:
- Pressure vessels in which pressure is exerted by gasses or vapours, liquids or solids with a gas or vapor cushion and liquids at a temperature above boiling point temperature at atmospheric pressure:
    - Group I:
        - Low temperature liquid gases with a permissible working pressure between 0.01 and 0.1 bars
        - Working pressure ≤ 25 bars and (Working pressure in bars x volume in litters) ≤ 200
        - Pipes with inside cross section ≤ 100 cm² and (Working pressure in bars x ID in mm) ≤ 2000

- Group II:
  - Working pressure > 25 bars and (Working pressure in bars x volume in litters) $\leq$ 200
  - Working pressure $\leq$ 1 bar and (Working pressure in bars x volume in litters) > 200
- Group III:
  - Working pressure > 1bar and (Working pressure in bars x volume in litters) is between 200 and 1000
- Group IV:
  - Working pressure > 1bar and (Working pressure in bars x volume in litters) > 1000

- Pressure vessels in which pressure is exerted only by liquid, the temperature of which does not exceed the boiling temperature at atmospheric temperature:
  - Group V:
    - Working pressure $\leq$ 500 bars
    - Working pressure > 500 bars and (Working pressure in bars x ID in mm) $\leq$ 1000
  - Group VI:
    - Working pressure > 500 bars and 1000 < (Working pressure in bars x volume in litters) $\leq$ 10000
  - Group VII:
    - Working pressure > 500 bars and (Working pressure in bars x volume in litters) > 10000

- Allowable design stress:

Allowable Design Stress = Yield Strength at Design Temperature / Safety factor

Safety factor = 1.5 for rolled plate, forged shell
= 2.0 for cast steel
= 1.5 for Aluminum and alloys
= 3.5 for cast iron

TEMA: Tubular Exchanger Manufacturers Association
- (TEMA) is trade association of leading manufacturers of shell and tube heat exchanger and is founded in 1939.

- **Scope :**
    - ID < 100 inches (2540 mm)
    - ID inches (mm) * Design pressure psi < 10000
    - Design pressure < 3000 psi

- **Classes of Heat Exchangers:**
    - Class 'R' Exchangers :
        - For severe requirements of petroleum and related processing applications.
    - Class 'C' Exchangers :
        - For moderate requirements of commercial and general processing applications.
    - Class 'B' Exchangers :
        - For chemical process service.
- **Construction Code : ASME Sec. VIII Div. 1**

ASME Section VIII Division-1, 2 & 3
- Historical Development of ASME Section VIII Div- 1, 2 & 3
    - In the early 20th century, explosion of steam boilers in U.S was frequent. Occurring rate 1/day.
    - 1914: ASME Boiler and pressure vessel code is published.
    - 1925: First publication of Section VIII – Unfired Pressure Vessels.
    - 1934: API+ASME jointly published unfired pressure vessel code for petroleum industry.
    - 1952: These two codes are merged in to single code: ASME Unfired Pressure Vessel Code Section VIII
    - 1968: ASME Section VIII Div-2 is published and original code became Div-1
    - 1997: ASME Sec VIII division 3 is published.

ASME Code Organization

- I   Rules for Construction of Power Boilers
- II   Materials
    - Part A – Ferrous Material Specifications
    - Part B - Nonferrous Material Specifications
    - Part C - Specifications for Welding Rods, Electrodes, and Filler Metals
    - Part D - Properties
- III Nuclear Power Plant Components
- IV   Rules for Construction of Heating Boilers
- V   Nondestructive Examinations
- VI   Recommended Rules for Care and Operation of Heating Boilers.
- VII Recommended Guidelines for Care of Power Boilers
- VIII Rules for Construction of Pressure Vessels.
    - Division 1: Rules for Construction of Pressure Vessels.
    - Division 2: Alternative Rules for Construction of Pressure Vessels.
    - Division 3: Alternative Rules for Construction of High Pressure Vessels.
- IX   Welding and Brazing Qualifications
- X   Fiber-Reinforced Plastic Pressure vessels
- XI   Rules for in-service inspection of Nuclear Power Plant Components

ASME Section VIII Division-1, 2 & 3

- ASME Section VIII codes are most widely used worldwide.
- Section VIII is divided into three divisions:
    - Division 1: Rules for construction of pressure vessels
    - Division 2: Alternative Rules for construction of pressure vessels.
    - Division 3: Alternative Rules for construction of high pressure vessels.
- Division 1 is used most often since it contains sufficient requirements for majority of pressure vessels.
- The main objective of ASME Code rules is to establish the minimum requirements that are necessary for safe construction and operation.
- ASME Code defines the requirements for material, design, fabrication, inspection and testing which are needed to achieve a safe design.

ASME Section VIII Division-1
- Scope:
    - Applicable for pressure between 15 psig and 3000 psig.
    - In relation to the geometry, scope of this division for the parts attached to pressure vessel is limited up to..
        - first circumferential joint for welded connection.
        - first threaded joint for screwed connection.
        - face of first flange for bolted flanged connection.
        - first sealing surface for proprietary connections or fittings.
    - Codes does not apply to non pressure parts, however welds which attach non pressure parts on pressure parts shall meet code rules.
    - Code identifies specific items for which it is not applicable. This includes:
        - Fired process tubular heaters (e.g. furnaces).
        - Pressure containers which are integral part of mechanical devices (e.g. pumps, turbines).
        - Piping system and their components.

- Structure of Sec VIII Div-1: It is divided into three subsections:
  - Subsection A: It consists part UG, general requirements that apply to all pressure vessel parts, regardless of material and fabrication method.
  - Subsection B: It covers requirements that apply to various fabrication methods. Subsection B consists of
    - Part UW for welded construction
    - Part UF for forged construction
    - Part UB for brazed construction
  - Subsection C: It covers requirements for several classes of material. Subsection C consists of-
    - Part UCS for carbon and low alloy steel material parts.
    - Part UNF for non ferrous material parts
    - Part UHA for non high alloy steel material parts
    - Part UNF for non ferrous material parts
    - Part UCL for clad and lined materials
    - Part UCD for cast and ductile parts
    - Part UHT for ferritic steel with properties enhanced by heat treatment
    - Part ULW for layered vessels
    - Part ULT for low temperature materials
- Division 1 also consists mandatory and non mandatory appendices:
  - Mandatory Appendices:
    - It addresses the subjects that are not covered elsewhere in the code.
    - Requirements of this appendices are mandatory.
  - Nonmandatory Appendices:
    - It provides information and suggested good practices.
    - Requirements of these appendices are not mandatory unless specified in purchase order.

ASME Section VIII Division-2
- Scope of division 2 is identical to that of division 1, however there is no limitation on higher pressure.
- Allowable stress values are higher for division 2 that division 1, hence division 2 vessels are thinner
- Vessels constructed with division 2 are economical over division 1 when saving in material cost is higher than the additional cost required to meet stringent requirements of division 2.
- Economical for higher pressure applications and for more expensive alloy materials.

ASME Section VIII Division-3
- Applied to vessels operating at internal or external design pressure is generally above 10000 psig.
- Design rules:
  - Maximum shear stress
  - Elastic plastic analysis
  - Fracture mechanics evaluation
- Material and NDE requirements more stringent than division 1 and 2

ASME Section VIII Division-1: Design
Following loadings to be considered in designing the vessels as per ASME Sec. VIII Div. I (clause UG-22)
- Internal design pressure at design temperature.
- External design pressure at design temperature.
- Weight of the vessel at operating and test condition.
- Load due to attached equipments.
- Weld joint efficiency.
- External / internal attachment
- Seismic loading.
- Wind loading.
- Fatigue loading due to thermal and pressure variations.
- Corrosion allowances.

It covers the design of pressure components as listed below
- UG-27 ----- Thickness of shells under internal pressure
- UG-28 ----- Thickness of shells under external pressure
- UG-29 -----Stiffening rings of cylindrical shells under external pressure
- UG-32 & 33 ---- Formed head , covers ellipsoidal, torisherical, hemisherical, conical & toriconical heads.
- UG-34 ---- Unstayed flat heads & covers
- UG-36 & 37 ---- Opening in pressure vessels & reinforcements
- UG-41 ---- Strength of reinforcements

It covers the design of pressure components as listed below
- UG -47 ----- Braced & stayed surface
- UG – 53 – Ligament
- UW-12 & 15 – weld efficiency & welded connection design
- UCS -56 – Requirement of PWHT
- UG -99 & 100 – Hydrotest & pneumatic test requirement
- Appendix – 2 – Design of bolted flange connection
- Appendix – 9 – Design of jacketed vessel
- Appendix -13 – Design of non-circular vessel
- Appendix – Y – Design of flat face flanges

Part	Thickness, t_p, in.	Pressure, P, psi	Stress, S, psi
Cylindrical shell	$\dfrac{Pr}{SE_1 - 0.6P}$	$\dfrac{SE_1 t}{r + 0.6t}$	$\dfrac{P(r + 0.6t)}{tE_1}$
Spherical shell	$\dfrac{Pr}{2SE_1 - 0.2P}$	$\dfrac{2SEt}{r + 0.2t}$	$\dfrac{P(r + 0.2t)}{2tE}$
2:1 Semi-Elliptical head	$\dfrac{PD}{2SE - 0.2P}$	$\dfrac{2SEt}{D + 0.2t}$	$\dfrac{P(D + 0.2t)}{2tE}$
Torispherical head with 6% knuckle	$\dfrac{0.885PL}{SE - 0.1P}$	$\dfrac{SEt}{0.885L + 0.1t}$	$\dfrac{P(0.885L + 0.1t)}{tE}$
Conical Section ($\alpha = 30°$)	$\dfrac{PD}{2\cos\alpha(SE - 0.6P)}$	$\dfrac{2SEt\cos\alpha}{D + 1.2t\cos\alpha}$	$\dfrac{P(D + 1.2t\cos\alpha)}{2tE\cos\alpha}$

Summary of ASME Code Equations

ASME Section VIII Division-1: Fabrication
It covers the fabrication requirements as listed below
- Weld joint categories as per UW-3
- Weld joint efficiency as per UW -12
- Welding of nozzles with shell as per UW -16
- Welding/welder qualification as per ASME Sec IX
- PWHT requirement as per UG-56
- Impact test requirement as per UCS-65 & UG 84
- Fabrication tolerances as per UG-80

Introduction - ASME Sec VIII Div 1

ASME SEC –VIII DIV- 1 Consists of:
Introduction – U
Subsection A - General Requirements - (Part UG)

Subsection B -Specific Requirements - Methods
UW – Welded construction
UF – Forging construction
UB – Brazing construction

Subsection C - Specific Requirements - Materials
UCS – Carbon steel & low alloy steel
UHA – High alloy steels
UCI – Cast Iron
UCD – Cast Ductile Iron
UCL – Cladded construction
ULW – Layered construction
UNF – Non ferrous metallurgy

Mandatory Appendices - Appendix 1 to 40

Non Mandatory Appendices - Appendix A, C, ... LL, MM (25 nos.)

Scope - ASME Sec VIII Div 1
U-1-(a)
Pressure vessels are containers for the containment of pressure, either internal or external. This pressure may be obtained from an external source, or by the application of heat from a direct or indirect source, or any combination thereof.

Excluded from ASME Sec VIII Div 1 scope
U-1 (C) (2) – (a,b,c) Vessels constructed as per other sections of the code, Tubular process heaters.
Components of pumps / compressors such as casings etc., Piping systems, Piping components like pipe, flange, fittings, strainers etc. & U-1 (C) (2) – (d,e,f,g

U-1 (C) (2) (h) (1) - Vessels having an internal or external operating pressure not exceeding 15 psi with no limitation on size.

U-1 (C) (2) (i) - Vessels with ID or height or width or cross section diagonal not exceeding 6" with no limitation on length of vessel or pressure.

U-1 (C) (2) (j) - Pressure vessels for human occupancy

U-1 (d) - Vessels with design pressure more than 3000 psi (210 bar)

Included in ASME Sec VIII Div 1 scope
See U-1 (e) to U-1 (e) (h)

Other Standards
BS 5500, IS 2825, Ad Merk Blatter (German), AFNOR (French), etc.

Material Identification

UG-4 to UG-8, UCS-5 – General & Materials
UCS-5 - Max. Carbon content - 0.35% for welded construction & for oxy. Cutting
UG-10 – Materials not permitted by this div. / not fully identified
UG-16 (C) - Mill under tol.- Smaller value of 0.25 mm or 6% of ordered thk.
UG-77,93,94 & - Material Identification
UG-116 – Marking / Stamping on vessel
UG-118 & UW-37 (f) – Stamping shall not be used of steel plates less than 6 mm thk. and of nonferrous plates less than 13 mm.
Table UCS-23 - List of material specifications for carbon & low alloy steels
Table UHA-23 - List of material specifications for High alloy steels
UCS-27 - Shells made from pipe
UCS-85 - Heat Treatment of test specimens (Simulation Heat Treatment)

Part of UCS

Table UCS - 23

Spec.No.	Type/Grade	Spec.No.	Type/Grade	Spec.No.	Type/Grade
SA-105...		SA-234	WPB, WPC, WPR, WPL, WP5, WP9, WP11 Cl 1.	SA-387	2,5,11,12,21,22,91
SA-106	A, B, C		WP12, Cl.1, WP22 Cl.1	SA-420	WPL 3, WPL. WPL 9
SA-179...		SA-266	1, 2, 3, 4	SA-508	1,1A,2Cl.1.2Cl.2,3 Cl.1,2 Cl.2.3V,4N Cl.
SA-182	FR, F1, F2, F3V, F5 F5a,F9,F11 Cl.1&2	SA-285	A, B, C		3, 22 Cl.3
	F12 Cl.1 & 2, F21	SA-302	A,B,C,D	SA-515	60, 65, 70
	F22 Cl. 1&3, F22V	SA-320	L7, L7A, L7M, L43	SA-516	55, 60, 65, 70
SA-193	B5, B7, B7M, B16	SA-333	1, 3, 4, 6, 7, 9	SA-533	A Cl.1&2 Cl.1&2,C Cl. 1&2, D Cl.2
SA-203	A, B, D, E, F	SA-335	P1,P2,P5,P5b,P5c,P9,P11 P12, P15, P21, P22,P91	SA-537	Cl. 1, 2 & 3
SA-204	A, B, C				
SA-209	T1, T1a, T1b	SA-336	F1,F3V,F5,F5A,F9,F11 Cl.2&3,F12,F21	SA-540	B21,B22,B23,B24,B24V
SA-213	T2,T5,T5b,T5c,T9 T11, T12, T17, T21 T 22, T91		Cl. 1&3, F22 Cl. 1 & 3, F22 V, P91	SA-541	1,1A,2Cl.1.2 Cl. 2,3 Cl.1, 3 Cl.2, 3V, 22 Cl. 3, 22V
		SA-350	LF1,LF2,LF3,LF5,LF9	SA-542	B Cl. 4, Cl. 4a, D Cl.4a

MATERIAL TEST COUPON – MTC. UCS-85

1. What is Material Test Coupon (MTC)?
A sample of actual raw material used in manufacturing main or critical vessel component.

2. What is the Purpose of MTC?
Raw material used for vessel components undergoes various heat treatment during manufacturing and final PWHT.
Each of these heat treatment alters mechanical properties of material.

MTC Facilitates simulation heat treatment & destructive testing of actual material representing vessel components for their mechanical properties as the final heat treated condition.

MATERIAL TEST COUPON - MTC

1. When MTC is Required ? UCS - 85

   UCS-85(c)- Whenever material undergoes any heat treatment during manufacturing above the lower transformation temperature (NR) and PWHT.

   UCS-85 (f) - P No.1, Gr.No.1 & 2 & all carbon & low alloy steels are exempted for MTC if heat treatment during fabrication is limited to post weld heat treatment at temperatures below the lower transformation temperature of the steel.

2. Extend of Testing Requirement
   - As Per Material Specification – ASME Sec. II Part A
3. Acceptability of Test Results
   - Minimum values specified in the material specification
4. MTC shall undergo all heat treatment including final PWHT along with the job prior to testing. Or
   - MTC can be subjected to a simulation heat treatment in advance by vessel manufacturer and tested prior to releasing the material for production.
   - Alternatively, vessel manufacturer can obtain the raw material from Mill duly an MTC tested with predetermined simulation heat treatment in advance. Vessel manufacturer shall provide total time and heating cycle to the Mill along with the purchase order.

Dish Ends

UG-32 – Formulas for CR, KR
UG-81 - Tol. For formed heads - Crowning measured from ID of head, Over crowning + 1.25% ID of vessel and Under crowning – 5/8 % ID of vessel.
UG-80 - Out of Roundness Tol. - The. Diff. Bet.max. & min. ID shall not exceed 1% of the nominal dia (ID/OD).

UCS-79 - Forming Shell sections & Heads - Heat treatment is required for all cold formed parts if fiber elon. is more than 5%, except for P No.1 Gr.No.1 & 2 materials may be as great as 40% with some conditions mentioned.

- **For P No.1 Gr.No.1 & 2 materials, If the nominal thickness is above 16 mm & fiber el. is more than 5%– SR is mandatory.**

UHA – 44 – Requirements for post fabrication heat treatment (solution annealing) due to straining for high alloy steels.

UW-34 - Spin Hole - Max. Dia - 60 mm - MPT/PT & RT required.

INSPECTION OF DISHED ENDS
TYPE OF DISHED ENDS -
- **Hemispherical**
- **Ellipsoidal**
- **Torispherical**

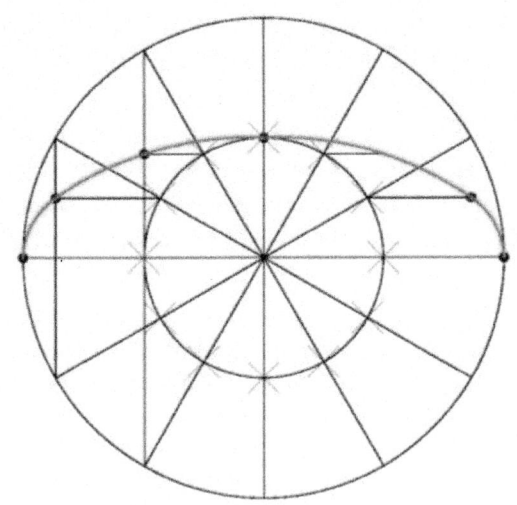

THE CONSTRUCTION OF ELLIPSE
Describe a circle of which diameter is equal to the major axis of the ellipse and with the same center a circle of which diameter is equal to the minor axis. Divide both the diameters in equal parts. From the intersecting points of the large circle draw perpendicular lines to the major axis and from the intersecting of the small circle draw lines parallel with the major axis. The intersections of these parallel and perpendicular lines are points of the elliptical curve.

INSPECTION OF DISHED ENDS..

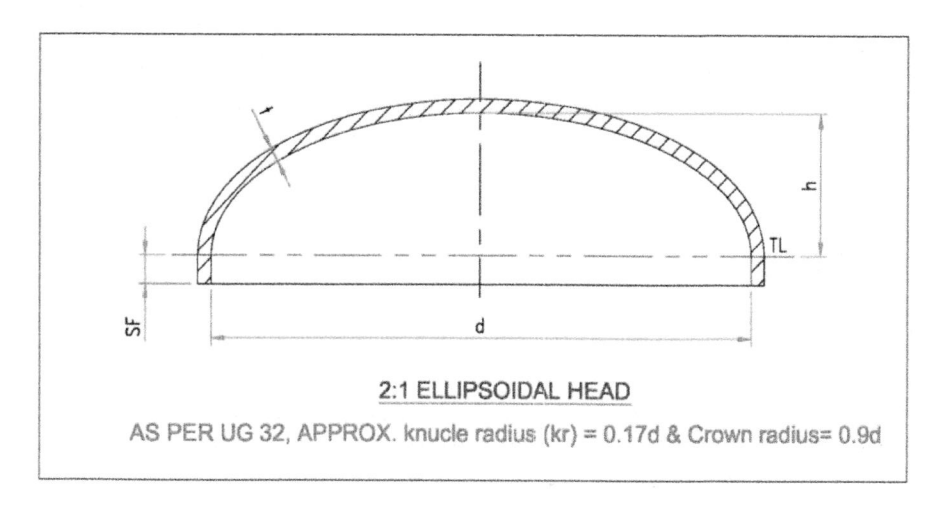

2:1 ELLIPSOIDAL HEAD

AS PER UG 32, APPROX. knucle radius (kr) = 0.17d & Crown radius = 0.9d

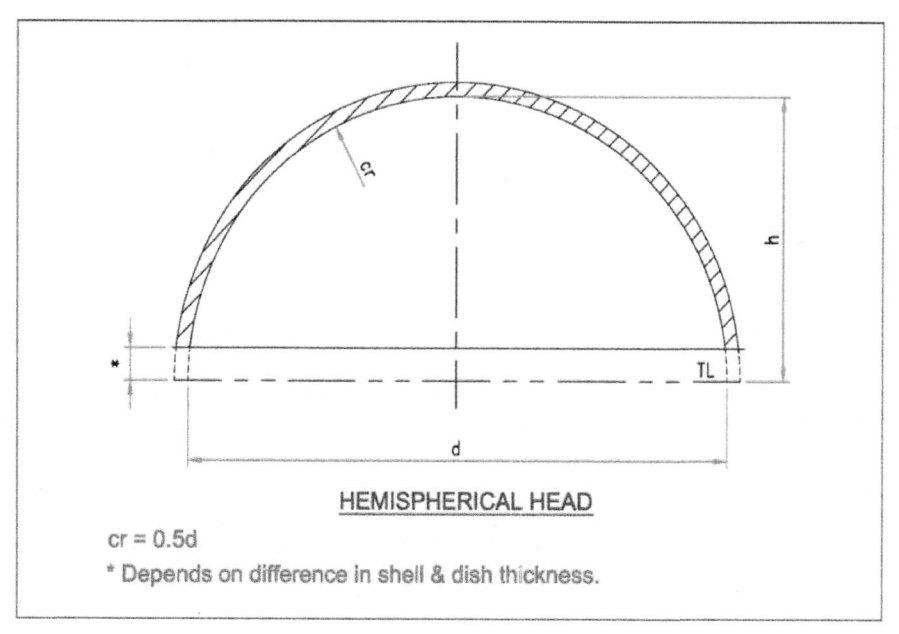

HEMISPHERICAL HEAD

cr = 0.5d

* Depends on difference in shell & dish thickness.

INSPECTION OF DISHED ENDS..

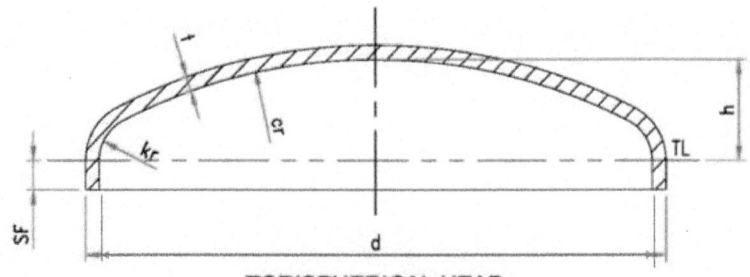

TORISPHERICAL HEAD

d= inside diameter, kr = inside knuckle radius, cr= inside crown radius, SF= straight flange, t=thk, TL= Tan line, h= inside depth

AS PER UG 32,
kr ≥ 6% of cr but not less than 3 times thickness.
cr ≤ Out side diameter of skirt.

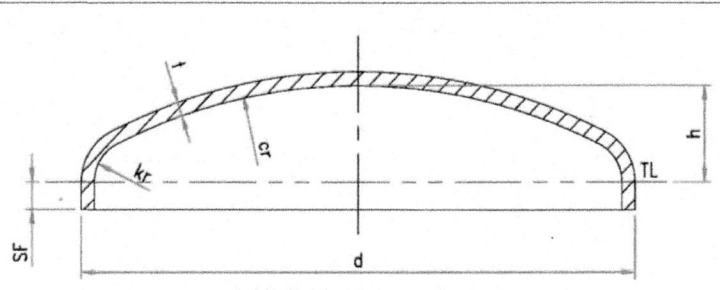

TORISPHERICAL HEAD

d= inside diameter, kr = inside knuckle radius, cr= inside crown radius, SF= straight flange, t=thk, TL= Tan line, h= inside depth

AS PER UG 32,
kr ≥ 6% of cr but not less than 3 times thickness.
cr ≤ Out side diameter of skirt.

INSPECTION OF FORMED DISHED ENDS

Applicable For Single Piece & Crown + Petal Construction

1. Circumference At Trimming line (Acceptance as per Tolerance Chart)
2. Knuckle Profile With 2D Full Size Template (No Gap At Center – End Gap max. 2mm
3. Height From Tan line (+ Half of 1¼ %D/E Ø, – Half Of 5/8 % D/E Ø)
4. Thickness At Crown, Knuckle & Edge (Above Minimum Specified In Drg.
5. Ovality On Open End (¼ Of 1% Of D/E Ø).
6. Check – Straight Face (Tan Line To Trimming Line) With A Straight Edge of 1 Foot Scale. Minimum As Specified In Drawing. (Not Applicable To Hemispherical D/E)
7. All Press Marks On Out Side & Inside Shall Be Smoothly Ground / Merged.
8. PT Check Out Side Surface – No Linier Indication Permitted
9. Check - Profile, Tan Line & Trimming Line With Full Size 2D Template. (Template radius – Open End, Crown & Knuckle - Less by 20 mm.) Template Shall Have Markings Of Tan Line, Trimming Line & Crown Centre. Deviation Permitted – Half of 1¼ %D/E Ø in Plus & Half Of 5/8% D/E Ø in Minus; No Abrupt Change In Profile. (Template To Be Verified On Lay Out)
10. Water Level 3 Points at 120° Apart On Trimming Line & Check Crown Center By A Plumb. (Deviation 5 mm per Meter Length Of D/E Ø Max. 20mm.(Not Applicable To Hemispherical D/E).
11. If Hot Formed, Check For MTC With Simulation Heat Treatment. (Not Applicable To Cold Formed D/E).
12. Blanks With Weld Joints Shall Be PT Checked & Radiographed Before & After Forming.

INSPECTION OF FORMED DISHED ENDS

Profile Check At Knuckle Portion

Gap At Centre
Not Acceptable

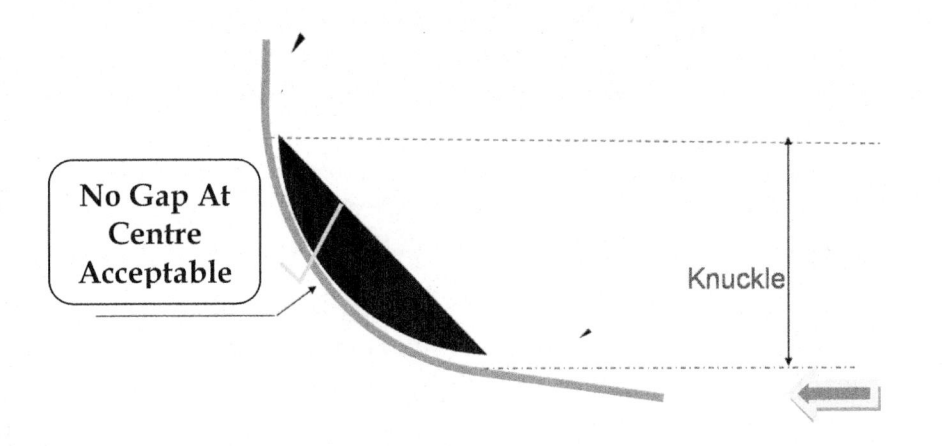

Knuckle

INSPECTION OF FORMED DISHED ENDS

Profile Check At Knuckle Portion

No Gap At Centre
Acceptable

Knuckle

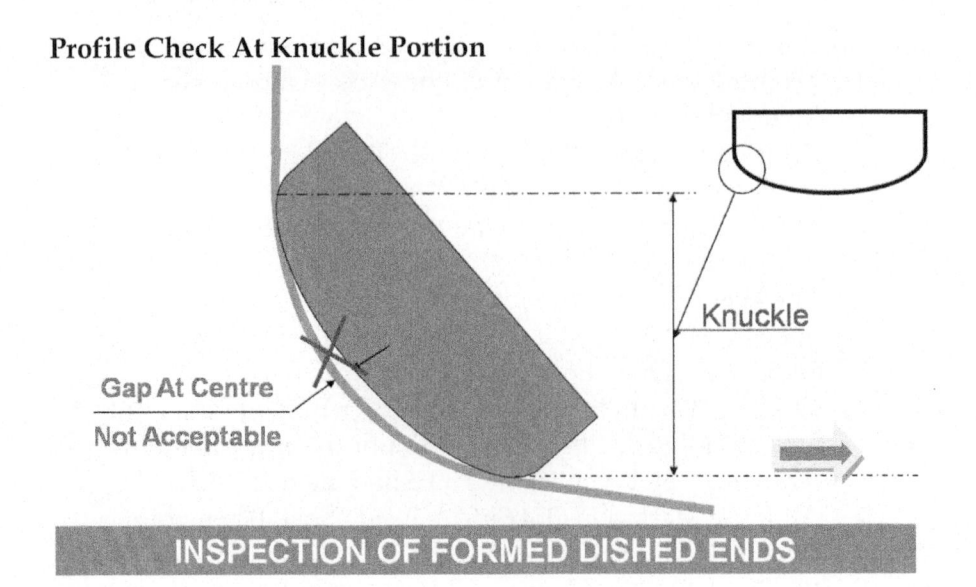

INSPECTION OF FORMED DISHED ENDS

2D Full Size Template Check. Profile-Tan Line-Trimming Line.

Repeat Marking & Checking Along the Full Circumference at Every 300 mm apart.

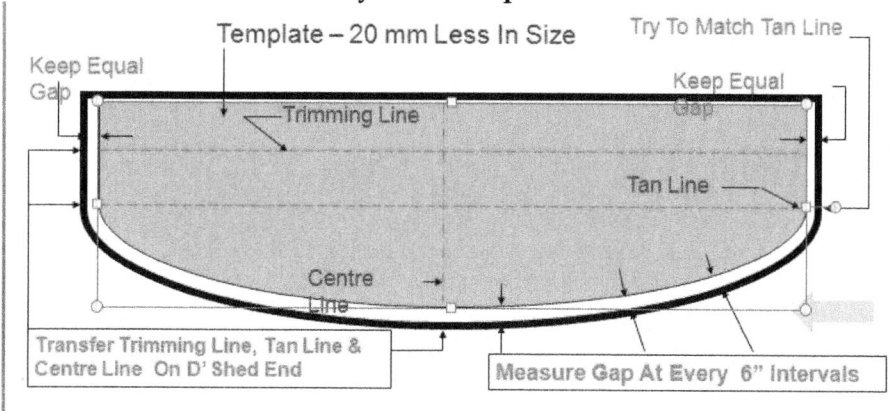

Template – 20 mm Less In Size
Try To Match Tan Line
Keep Equal Gap
Keep Equal Gap
Trimming Line
Tan Line
Centre Line
Transfer Trimming Line, Tan Line & Centre Line On D' Shed End
Measure Gap At Every 6" Intervals

INSPECTION OF FORMED DISHED ENDS
Applicable For Only Crown + Petal Construction

1. Each Formed Petal Segment Shall Be Checked By A 3D Template From Inside. Profile Of Template Ribs To Be Verified. Maximum Gap Permissible At Any Location At The Rate Of 3mm / Mtr. Length Of Petal With Maximum Of 6mm.
2. Crown Segment Shall Be Checked By 2D Template. Gap Permissible 3mm / Mtr. Blank Dia.
3. **Checks After Assembly -**
    - Off Set – Half Of Permissible As Per Code.
    - Root Gap – (– 1mm , + 2mm)
    - Root Face – (± 1mm)
    - Serrations If Any On W E P, To Be Ground Smooth
    - No Temporary Tacks On Groove
    - No Defects On Tacks With External Cleats
    - Petals To Be Numbered With D/E No.
    - Match Marks On Adjacent Petals

MARKING OF CENTRE LINES ON DISHED ENDS

1) KEEP D/E ON THREE SUPPORTS
2) WATER LEVEL TAN LINE OR TRIMMING LINE
3) MEASURE CIRCUMFERENCE AND DEVIDE IN TO 4 EQUAL PARTS
4) MARK 0°, 90°, 180° & 270° ORIENTATIONS
5) CLAMP VERTICAL STRAIGHT EDGES AT ALL ORIENTATIONS. VERTICALITY TO BE VERIFIED WITH SPIRIT LEVEL
6) STRETCH TWO TWINES DIMETRICALLY OPPOSITE ALONG THE STRAIGHT EDGES AND BRING DOWN TO MARKED ORIENTATIONS
7) MARKS THE LOCATION OF TWINE (CENTRE LINES) ON THE D/E SURFACE

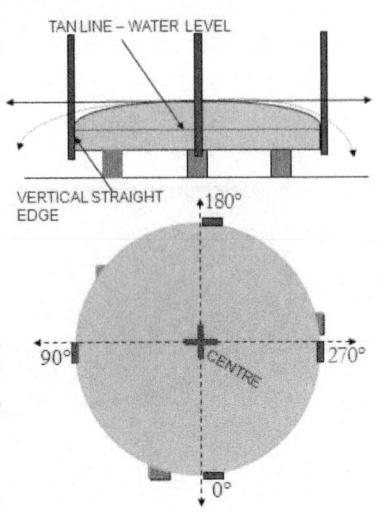

Openings

UG-36 (b) (1) – Maximum Size of Opening in a vessel

UG-36 (c) (3) – Size of Opening in a vessel which will not require reinforcement.

UG-37 (g) – Telltale hole requirement.

UG-42 (a) & (b) – Requirements for Reinforcement overlapping or combined reinforcement for two and/or more openings.

UW-14 – Openings in or adjacent to welds
- Any type of reinforced/strengthened opening can be located in the weld.
- Additional RT requirement for unreinforced Openings in or adjacent to welds.

Preheating

Nonmandatory Appendix R – Preheating
For P No.1, Gr.No.1, 2, 3 - 79 deg C for material which have both max. CE – 0.30 % and thk. at joint in excess of 25 mm and 10 deg C for all other materials in this P No.

UW-30 - Lowest permissible temp. for welding - -20 deg C (0 deg Fer), Preheat 15 deg C - within 75 mm area one side.

Other suggested Preheat temperatures –
Annex XI of AWS D 1.1 – Table 3.2
ANSI/ASME B31.3

Welding

UW-3 - Weld Joint catagories (Cat A- Long seam, B- Cir.Seam, C Joint connecting Nozzle to flanges, D- Joints connecting nozzle to main shell).

UW-9 - Weld Joint for different thknesses (Applicable only when thk differ more than 3 mm) – tappered transition length not less than 3 times the offset bet two surfaces.

    - Difference beween two long seams of adjacent courses atleast 5 times the thk. of thicker plate.

    - Lap joints - overlap shall not be less than 4 times the thk. Of inner plate.

UW-16 & UW-16.1
    - Reinforcement - Weld at outer edge of rein. plate shall be conti.fillet type & min. throat dim. 1/2 t min.

    - Reinforcement - Weld at inner edge of rein. plate shall be conti.fillet type & min. throat dim. 1/2 t min (If not abutting a nozzle).

    - Reinforcement - Weld at inner edge of rein. plate shall be conti.fillet type & min. throat dim. 0.7 t (abutting a nozzle).

    - Throat dimension = 0.7 X Leg length of equal fillet (ex. If equal fillet leg size 10 mm, then throat dimension will be 7 mm).

UW-21 - ASME B 16.5 Socket weld flanges - External fillet weld - min. filet weld throat dim. should be lesser of Nozzle wall thk or 0.7 times the hub thk.

UW-30 - Lowest permissible temp. for welding: -20 deg C (0 deg Fer), Preheat 15 deg C within 75 mm area one side

UW-33- Alignment tol. - Up to 13 mm thk - Max.offset - 1/4 t for all cat. Welds, Over 13 to 19 mm thk. – For Cat.A- 3 mm, for other cat. - 1/4 t.

UW-35 - Weld reinforcement detail - Thk. Over 4.8 to 13 - For B&C joints - Max. 4 mm, For other joints
    -2.4 mm, Thk. over 13 to 25 - For B&C joints - Max. 4.8 mm, For other joints -2.4 mm.

UCS-56 (f)- Weld repair allowed after PWHT

Part UW

- **UW-2 - Service Restrictions**

(a) Lethal Service
(b) Low Temperature Service
(c) Steam Service
(d) Direct fired Service

(a) Lethal Service

- Vessels with lethal substances, all butt welds shall be full radiographed.
CS/LAS vessels shall be post weld heat treated.
- It shall be responsibility of user to specify – if vessel is for lethal service.
- All "category A" joints shall be "type (1)"
- All "category B & C" joints shall be "type (1) or (2)"
- All "category D" shall be full penetration welds.

(b) Low Temperature Service

- All "category A" joints shall be "type (1)"
- All "category B" joints shall be "type (1) or (2)"
- All "category C" joints shall be full penetration welds.
- All "category D" joints shall be full penetration welds.

c) Unfired Steam Boilers
- All "category A" shall be "type (1)" and "B" shall be "type (1) or (2)".
- All butt welds shall be radiographed
- CS/LAS vessels shall have PWHT.

(d) Direct Fired Vessels
- All "category A" shall be "type (1)" and "B" above 16mm thickness "type (1) or (2)"
- CS above 16mm and LAS vessels shall have PWHT.
- Parts with other than type (1) & (2) welds subjected to radiation. Design temperature shall be maximum surface metal temp.

WELD JOINT CATEGORY – UW – 3

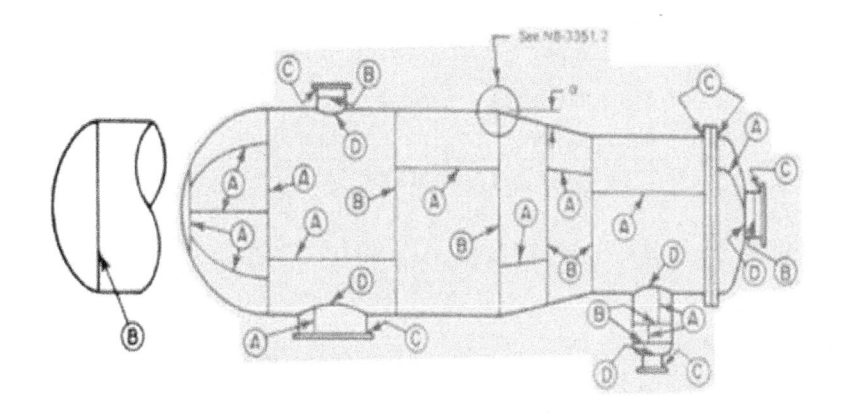

A = All Long Seams in Shells, Cones and Pipes;- All joints in formed D/Es or Flat Heads; - Circ. Seam Connecting Hemispherical D/Es To Shells / Pipes.

B = All Circ. Seams in Shells, Cones and Pipes. - Circ. Seam Connecting Shells / Pipes To D/Es other than Hemisphere.

C = Weld Joints Connecting Flanges/TS/ Flat Heads To Shells / Pipes /Formed D/Es.

D = Weld Joints Connecting Nozzles or Nozzle Pipes To Shells / Pipes/Heads.

MAXIMUM OFF SET IN BUTT WELD – UW 33

Joint Thickness ."t"	Long Seam (Category "A")	Circ. Seam (Category "B","C","D")
13 mm & Below	$1/4^{th}$ t	$1/4^{th}$ t
Over 13 mm to 19 mm	3 mm	$1/4^{th}$ t
Over 19 mm to 38 mm	3 mm	5 mm
Over 38 mm to 51 mm	3 mm	$1/8^{th}$ t
Over 51 mm	$1/16^{th}$ t. Max. 10 mm	$1/8^{th}$ t. Max. 19mm

Aim For Half Of The Permissible Off Set At Set Up Stage

REINFORCEMENT LIMIT ON BUTT WELDS - UW 35

Material Thickness	Max. Reinforcement	
	Long Seam (Category "A"&"D")	Circ. Seam (Category "B"& "C")
Below 2.4 mm	0.8 mm	2.4 mm
2.4 to 4.8 mm incl	1.6 mm	3.2 mm
Over 4.8 to 13 mm incl	2.4 mm	4.0 mm
Over 13 to 25 mm incl	2.4 mm	4.8 mm

REINFORCEMENT LIMIT ON BUTT WELDS Con..- UW 35

Material Thickness	Max. Reinforcement	
	Long Seam (Category "A"&"D")	Circ. Seam (Category "B"& "C")
Over 25 to 51 mm incl	3.2 mm	4.8 mm
Over 51 to 76 mm incl	4.0 mm	6.0 mm
Over 76 to 127 mm incl	6.0 mm	6.0 mm
Over 127 mm	8.0 mm	8.0 mm

Part UW (Continued..)

- **UW-13 Attachment Detail**
- **Heads Attached to Shells (continued..)**

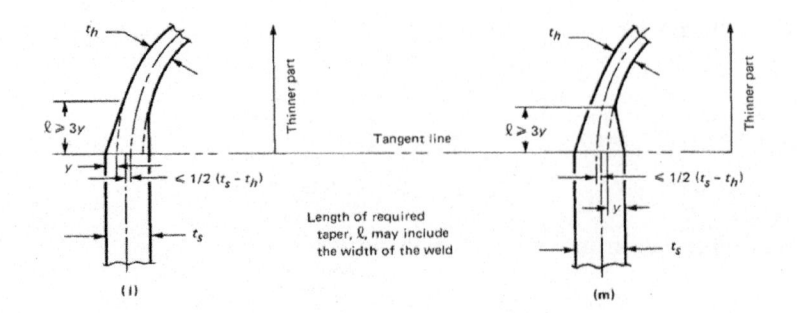

(i) $\ell \geq 3y$, $\leq 1/2\,(t_s - t_h)$

Length of required taper, ℓ, may include the width of the weld

(m) $\ell \geq 3y$, $\leq 1/2\,(t_s - t_h)$

In all cases, the projected length of taper ℓ shall be not less than $3y$. The shell plate center line may be on either side of the head plate center line.

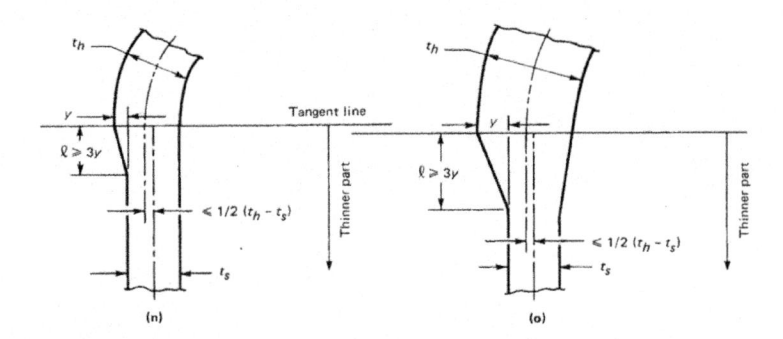

(n) $\ell \geq 3y$, $\leq 1/2\,(t_h - t_s)$

(o) $\ell \geq 3y$, $\leq 1/2\,(t_h - t_s)$

In all cases ℓ shall not be less than $3y$ when t_h exceeds t_s. Minimum length of skirt is $3t_h$ but need not exceed 1-1/2 in. except when necessary to provide required length of taper.
When t_h is equal to or less than $1.25\,t_s$, the length of skirt shall be sufficient for any required taper.

Length of required taper ℓ may include the width of the weld. The shell plate center line may be on either side of the head plate center line.

Part UW
- ## UW-16   Nozzle welded to shells

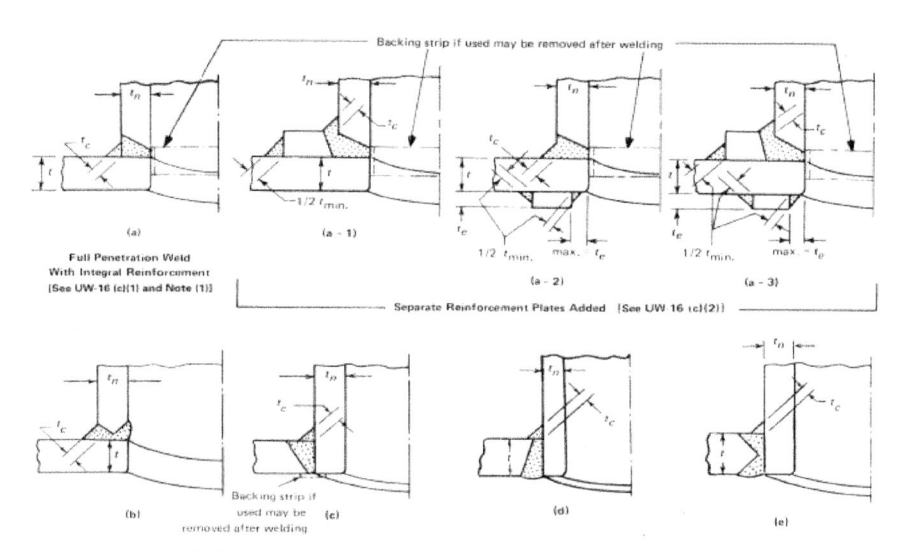

FIG. UW-16.1 SOME ACCEPTABLE TYPES OF WELDED NOZZLES AND OTHER CONNECTIONS TO SHELLS, HEADS, ETC.

- ## UW-16   Nozzle welded to shells (continued)

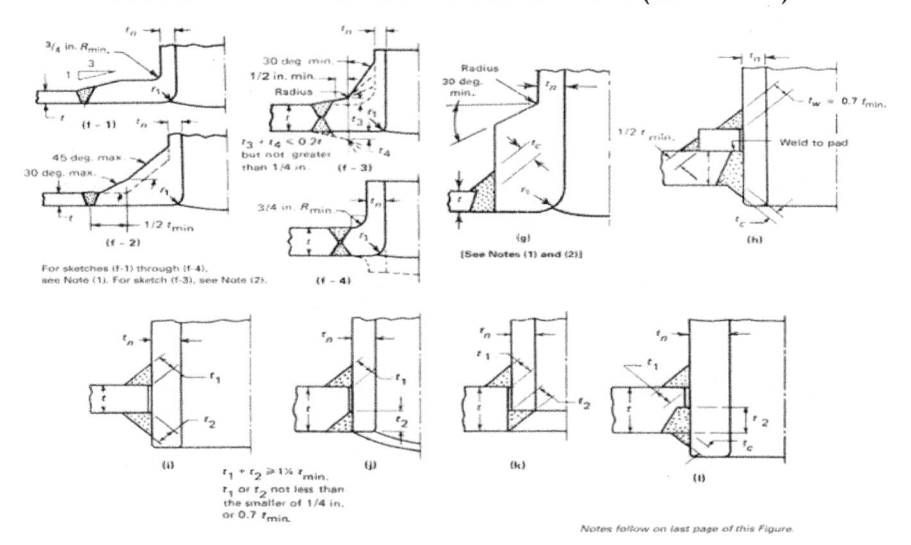

FIG. UW-16.1 SOME ACCEPTABLE TYPES OF WELDED NOZZLES AND OTHER CONNECTIONS TO SHELLS, HEADS, ETC. (CONT'D)

RT

UW-12 - Types of weld, joint efficiency & RT requirements.

UW-11- RT & UT - for butt weld joints.

RT for All but weld (Full Radiography) for vessel containing lethal substances, thk. 38 mm & over for all materials, For P No.1 full RT above 32 mm thk. As per UCS-57.

UW-11(a) - (5) (b) – Spot Radiography for T joints

Cat.B & C joints less than NPS 10 & wall thk. less than 29 mm - do not require RT

UT can be done instead of RT where interpretable radiographs can not be achieved.

UCS-57- Mandatory requirement of full RT for each butt weld joint above certain thks for all materials (Table for all P Nos. of CS and AS).

UW-51- RT of Welded Joints, RT as per Section 5, Written RT procedure is not required

UW-14 – Additional RT requirement for Openings in or adjacent to welds.

UW-51- Acc. Crit – Full Radiography
Elongated Indications:
(1) Crack, IF,IP not allowed **(2)** Max. 6 mm up to 19 mm thk. **(3)** max. 1/3 t for thk. 19 mm to 57 mm. **(4)** Alligned indication. **- Not acc. When aggregate length greater than t in a length of 12t**, except they are seperated by more than a distance of 6L, where L is the length of longest imperfaction.

RT

Mandatory Appendix 4 - Rounded Indications chart acc. criteria –
RT of welds, Rounded indication- with a Max. length of three times the width or less Aligned indications - A sequence of four or more rounded indications in a parallel line.
Aligned indications are acceptable when the sum of the dia of the indications is less than t in a length of 12t.
And also the spacing bet'n the groups of aligned indications shall meet the req. as shown in figure 4-2.
Relevant Indication - As mentioned in 4-3
Max. size of rounded indication - 1/4 t or 4 mm, whichever is smaller except that an isolated indication separated from an adjacent indication by 25 mm or more may be 6 mm.
Max. size of acceptable rounded indication (Random & Isolated) - according to the job thk. As per table 4-1.
Clustered indications - The length of an acceptable cluster shall not exceed the lesser of 25 mm or 2 T.
When more than one cluster is present, the sum of the length of clusters shall not exceed 25 mm in a 150 mm length weld.

UW-52 - Spot Radiography
One spot o 6 inch per 15 m of weld, spot selection by inspector only, if welder changes, additional spot RT is required. Acc. Crit-
(1) Crack, IF,IP not allowed
(2) **Slag/cavity - Max. length of indication 2/3 t**
(3) For all thk - indications less than 6 mm are acceptable.
(4) (4) For all thk - indications greater than 19 mm are unacceptable.
(5) **Alligned indi. - Not acc. when aggregate length greater than t in a length of 6t,** except they are seperated by more than a distance of 3L, where L is the length of longest imperfaction.
(6) **Rounded Indications are not a factor in the acceptability of welds not required to be fully radiographed.**

If RT is failed Spot exam. Two additional spot RT is to be performed on the same weld increment.

If RT is failed as mentioned above in retest, the entire weld increment shall be rejected, removed & re welded & shall be completely radio graphed or at the fabricator's option, the entire increment of weld represented shall be completely radiographed and only defects need to be Corrected.

UW-52 (b) (4) - T joints are excluded from spot radiography.

Part UW (Continued..)

- **UW-11 - Radiographic Examination**

(a) Full Radiography
(1) All butt welds in shell & heads of vessels with lethal service.
(2) All butt welds, where t>38 mm or exceeds lesser thickness specified elsewhere in code.
- Categories B and C butt welds in nozzles that neither exceed 10" size nor 29 mm thickness, do not require radiography.

All category A and D butt welds with joint efficiency [of full radiography]
(a) Cat. A and B butt welds of vessel sections or heads shall be of type (1) or (2).
(b) Cat. B or C butt welds (not including those in nozzles except as required above) which intersect Cat. A butt welds or connect seamless vessel sections or heads shall meet spot radiography requirement.

UT may be substituted for closing seam if RT is not possible.

- **UW-11 - Radiographic Examination**

(b) Spot Radiography
- Butt weld joints of type (1) or (2) which are not required to be fully radiographed.
-If spot radiography is specified for entire vessel, radiography is not required of Cat B and C butt welds in nozzles that exceed neither 10" size not 29 mm thickness.

(c) No Radiography
- No radiography is required when vessel is designed for external pressure only or where joint efficiency is as per table UW-12 (C).

- **UW-12 - Joint Efficiency**

- Values of E given in column (a) for fully radiographed butt joints.
- When UW-11(a)(5) is not met, value of E will be as given in (b)
- Value if E given in column (b) for spot radiographed butt joints.
- Value if E given in column (c) for no radiography joints.
- Seamless vessel sections/heads
E = 1.0 when spot radiography requirements of UW-11(a)(5)(b) are met.
E = 0.85 when above is not met, or when Cat A or B welds are type 3, 4, 5 or 6.

Table UW-12

Maximum Allowable Joint Efficiencies for Arc and Gas welded joints

Type No	Joint Description	Limitations		Joint Category	Degree of Radiographic Examination		
					(a) Full	(b) Spot	(c) None
(1)	Butt joints as attained by double-welding or by other means (backing strips which remain in place are excluded)	None		A, B, C & D	1.00	0.85	0.70
(2)	Single-welded butt joint with backing strip	(a)	None	A, B, C & D	0.90	0.80	0.65
		(b)	Circumferential butt joints with one plate offset	A, B, & C	0.90	0.80	0.65
(3)	Single-welded butt joint without use of backing strip	Circumferential butt joints only < 16 mm and < 600 mm OD		A, B, & C	NA	NA	0.60
(4)	Double full fillet lap joint	(a)	Longitudinal joints < 10 mm	A	NA	NA	0.55
		(b)	Circumferential joints < 16 mm	B & C	NA	NA	0.55

Part UCS

- **UCS-57 - Radiographic Examination**

-when thickness exceeds given in table, radiography is required.
-Thickness of thinner plate is considered.

TABLE UCS-57
THICKNESS ABOVE WHICH FULL RADIOGRAPHIC EXAMINATION OF BUTT WELDED JOINTS IS MANDATORY

P-No. & Gr. No. Classification of Material	Nominal Thickness Above Which Butt Welded Joints Shall Be Fully Radiographed, In.
1 Gr. 1, 2, 3	$1\frac{1}{4}$
3 Gr. 1, 2, 3	$\frac{3}{4}$
4 Gr. 1, 2	$\frac{5}{8}$
5A, 5B Gr. 1	0
9A Gr. 1	$\frac{5}{8}$
9B Gr. 1	$\frac{5}{8}$
10A Gr. 1	$\frac{3}{4}$
10B Gr. 2	$\frac{5}{8}$
10C Gr. 1	$\frac{5}{8}$
10F Gr. 6	$\frac{3}{4}$

PWHT

UW-40 - Procedure - PWHT

Soak Band shall include - Weld, HAZ & Base metal - Min. Width - Width of weld + (1t or 50 mm whichever is less) on each side of the weld.

Heating the vessel internally - int. press. Should not exceed 50% of working pressure

PWHT shall be performed before Hydrostatic test and after any weld repair except as permitted by UCS-56(f).

UCS-56 - Requirements – PWHT

PWHT requirements for different P Nos. as per table
For P No.1 - Mandatory if weld joint thk is over 38 mm and For welded joints over 1.25" nom. Thickness through 1.5" nom. Thickness, unless preheat is applied at a min. Temperature of 200'F (94'c) during welding.
Mandatory for lethal services
For P No.1 materials- Min. Hold. temp - 595 'C, Min. Holding time- 1 Hr/in. for thk. Up to 2" -15 min. minimum.

For P No.8 –PWHT is neither required nor prohibited.

UCS-56 (f)-Weld repair allowed after PWHT

UCS-68 - PWHT is Mandatory for all materials if MDMT is colder than -48 deg C and coincident ratio(as per UCS 66.1) is 0.35 or greater - Except this req. does not apply for P no.1 if impact test is done at MDMT - required value 34 J.

PWHT
1) The temperature of furnace shall not exceed 800'F (425 'C) at the time when the vessel or part is placed in it.
2) Above 800 'F(425 'C), the rate of heating shall not be more than 400 'F Per hour (222 'C/Hour) divided by the maximum metal thickness of the shell or head plate in inches, but in no case more than 400 'F Per hour(222 'C Per hour).
3) During the heating period, There shall not be a greater variation in temperature throughout the portion of the vessel being heat treated than 250 'F(140 'C) within any 15 feet (4.6m) interval of length.
4) During the holding period, there shall not be a greater difference than 150 'F (83 'C) between the highest and the lowest temperature the portion of the vessel being heated
5) Above 800 'F (425 'C), The rate of cooling shall not be more than 500 'F Per hour (280 'C/Hour) divided by the maximum metal thickness of the shell or head plate in inches, but in no case more than 500 'F Per hour (280 'C Per hour).
6) The rate of heating & cooling need not be less than 100 'F/Hr (56 'C/Hr).

POST WELD HEAT TREATMENT

- **UCS-56 – POST WELD HEAT TREATMENT (PWHT):**

- **OBJETCIVE:**
- TO REDUCE THE RESIDUAL STRESSES AT WELD & HAZ.
- TO MODIFY THE MICRO-STRUCTURE (HAZ)

- **EFFECT:**
- SUSCETIBLITIES TO BRITTLE FAILURE.

- **DECREASES.**
- RESISTANCE TO STRESS CORROSION CRACKING INCREASES.
- DIMENSIONAL STABILITIES INCREASES.

- **PWHT PARAMETER**
- HEATING RATE
- SOAKING TEMPERATURE
- COOLING RATE

Part UCS

- **UCS-56 – PWHT**

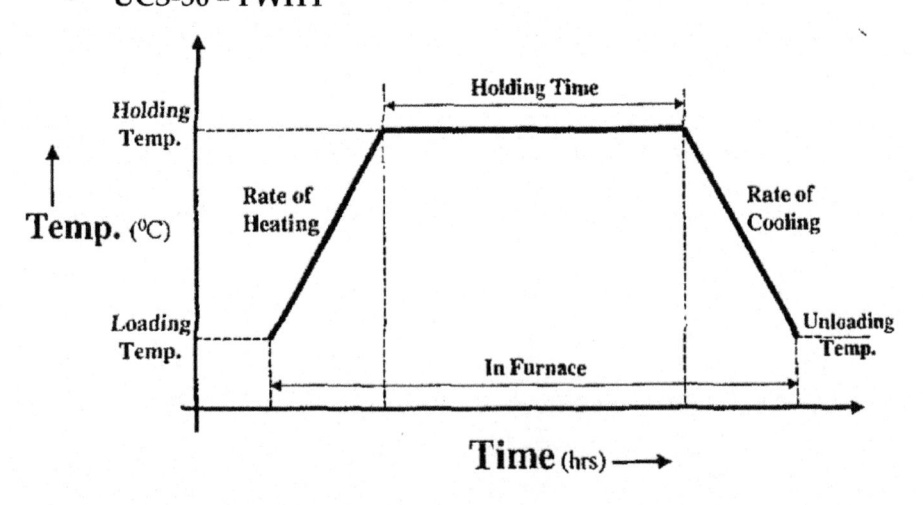

- **UCS-56 – PWHT**

-General
- Rate of heating and cooling need not be less than 100°F(38°C)/hr
- Holding temp./holding time in excess of min. value may be used.
- Intermediate PWHT need not confirm to table UCS-56
- Holding time need not be continuous. It may be addition of time of multiple PWHT cycles.
- When P numbers are different, higher PWHT temperature shall be used.
- When non pressure part is welded to pressure part, PWHT of pressure part shall control.
- Exemptions given are not permitted when PWHT is a service requirement.

- **Table UCS-56**

TABLE UCS-56
POSTWELD HEAT TREATMENT REQUIREMENTS FOR CARBON AND LOW ALLOY STEELS

Material	Normal Holding Temperature, °F, Minimum	Minimum Holding Time at Normal Temperature for Nominal Thickness [See UW-40(f)]		
		Up to 2 in.	Over 2 in. to 5 in.	Over 5 in.
P-No. 1 Gr. Nos. 1, 2, 3	1100	1 hr/in., 15 min minimum	2 hr plus 15 min for each additional inch over 2 in.	2 hr plus 15 min for each additional inch over 2 in.
Gr. No. 4	NA	None	None	None

NOTES:
(1) When it is impractical to postweld heat treat at the temperature specified in this Table, it is permissible to carry out the postweld heat treatment at lower temperatures for longer periods of time in accordance with Table UCS-56.1.
(2) Postweld heat treatment is mandatory under the following conditions:
 (a) for welded joints over 1½ in. nominal thickness
 (b) for welded joints over 1¼ in. nominal thickness through 1½ in. nominal thickness unless preheat is applied at a minimum temperature of 200°F during welding
 (c) for welded joints of all thicknesses if required by UW-2, except postweld heat treatment is not mandatory under the conditions specified below:
  (1) for groove welds not over ½ in. size and fillet welds with a throat not over ½ in. that attach nozzle connections that have a finished inside diameter not greater than 2 in., provided the connections do not form ligaments that require an increase in shell or head thickness, and preheat to a minimum temperature of 200°F is applied;
  (2) for groove welds not over ½ in. size or fillet welds with a throat thickness of ½ in. or less that attach tubes to a tubesheet when the tube diameter does not exceed 2 in. A preheat of 200°F must be applied when the carbon content of the tubesheet exceeds 0.22%.
  (3) for groove welds not over ½ in. size or fillet welds with a throat thickness of ½ in. or less used for attaching nonpressure parts to pressure parts provided preheat to a minimum temperature of 200°F is applied when the thickness of the pressure part exceeds 1¼ in.;
  (4) for studs welded to pressure parts provided preheat to a minimum temperature of 200°F is applied when the thickness of the pressure part exceeds 1¼ in.;
  (5) for corrosion resistant weld metal overly cladding or for welds attaching corrosion resistant applied lining (see UCL-34) provided preheat to a minimum temperature of 200°F is maintained during application of the first layer when the thickness of the pressure part exceeds 1¼ in.

NA = not applicable

CODE EXTRACT FOR HEAT TREATMENT

P. NO.	HOLDING TEMP.	NOM. THICKNESS	SOAKING PERIOD
1 (CARBON STEEL) & 3 (LOW ALLOY STEEL)	1100 DEG. F (593° C)	UPTO 2"	1 HR. PER INCH., HOWEVER 15 MINUTES MINIMUM
		OVER 2" TO 5"	2 HOURS, PLUS 15 MIN. FOR EACH ADDITIONAL INCH ABOVE 2"
		OVER 5"	2 HOURS, PLUS 15 MIN. FOR EACH ADDITIONAL INCH ABOVE 2"

*POST WELD HEAT TREATMENT IS MANDATORY ON P-NO.3 GR. NO. 3 MATERIAL IN ALL THICKNESSES.

Code Requirements for PWHT

Material	Normal Holding Temperature °C	Minimum holding time at normal temperature for nominal thickness		
		Up to 2"	Over 2" to 5"	Over 5"
P - No.1, P - No.3 [e.g. SA516GR60/70]	595	1hr / in. 15 min minimum	2 hr plus 15 min for each additional inch over 2"	2 hr plus 15 min for each additional inch over 2"
P - No.4 [e.g. SA387GR11]	650	1hr / in. 15 min minimum	1 hr / inch	5 hr plus 15 min for each additional inch over 5"
P - No.5A P - No.5B P - No.5C [Group no.1] [e.g. SA387GR22/ SA540Ty.D Cl.4a]	675	1hr / in. 15 min minimum	1 hr / inch	5 hr plus 15 min for each additional inch over 5"

Material	Normal Holding Temperature °C	Minimum holding time at normal temperature for nominal thickness		
		Up to 2"	Over 2" to 5"	Over 5"
P - No.5B Group no.2 [e.g.SA387GR91]	705	1hr / in. 15 min minimum	1 hr / inch	5 hr plus 15 min for each additional inch over 5"

- **PWHT Requirements TABLE UCS-56**

POSTWELD HEAT TREATMENT REQUIREMENTS FOR CARBON AND LOW ALLOY STEELS (CONT'D)

Material	Normal Holding Temperature °F (°C) Minimum	Minimum Holding Time at Normal Temperature for Nominal Thickness [See UW-40(f)]
P-No. 9A Gr. No. 1	1100 (595)	1 hr. minimum, plus 15 min./in. (25 mm) for thickness over 1 in. (25 mm)
P-No. 9B Gr. No. 1	1100 (595)	1 hr. minimum, plus 15 min/in. (25 mm) for thickness over 1 in. (25 mm)
P-No.10A Gr. No. 1	1100 (595)	1 hr. minimum, plus 15 min./in. (25 mm) for thickness over 1 in. (25 mm)
P-No.10B Gr. No. 1	1200 (650)	1 hr. minimum, plus 15 min./in. (25 mm) for thickness over 1 in. (25 mm)
P-No.10C Gr. No. 1	1000 (540)	1 hr. minimum, plus 15 min/in. (25 mm) for thickness over 1 in. (25 mm)
P-No. 10F Gr. No.1	1000 (540)	1 hr. minimum, plus 15 min./in (25 mm) for thickness over 1 in. (25 mm)

- Table UCS-56.1

TABLE UCS-56.1
ALTERNATIVE POSTWELD HEAT TREATMENT REQUIREMENTS FOR CARBON AND LOW ALLOY STEELS
Applicable Only When Permitted in Table UCS-56

Decrease In Temperature Below Minimum Specified Temperature, °F	Minimum Holding Time [Note (1)] at Decreased Temperature, hr	Notes
50	2	...
100	4	...
150	10	(2)
200	20	(2)

NOTES:
(1) Minimum holding time for 1 in. thickness or less. Add 15 minutes per inch of thickness for thicknesses greater than 1 in.
(2) These lower postweld heat treatment temperatures permitted only for P-No. 1 Gr. Nos. 1 and 2 materials.

Impact Test of Base metal / Weld / PTC

UCS-66 - Impact Test Exemption Curves & Table.

UG-20 (f) - Impact testing exemptions for P No.1, Group No.1,2 materials only.

UCS-67 & UG-84 (h) - Impact Tests of Welding Procedures.

UG-84 - Charpy Impact Tests.

UG-84 (c) (4) & Fig. UG 84.1 - Min.Energy Requirements for impact test.

UG-84 (d) (1) - Impact test of material is acceptable if it is mentioned in mfg. Tc

Table UG 84.3 - List of specifications for impact tested materials (SA-Plates-5,20,480,Pipe-333,Tube-334,Forging-350,Casting-352,Bolting-320,Pipe fitting-420).

UG-84 (f) - Impact testing of weld & HAZ.

UG-84 (g) (5) – Impact value for weld & HAZ shall be as high as those required for the base materials.

UG-84 (h) (2) – impact value to be considered for higher thk if two diff. thk. are joined.

UG-84 (i) (2) – Must requirement for PTC if WPS is made with impact test requirement

Impact Test of Base metal / Weld / PTC

UG-84 (f) (2) - After HT only if HT is applicable.

UG-84 (g) (1) - One face of test sample shall be within 1.5 mm from surface of test piece Test temp. should be same as per requirement of base material.

UG-84 (i) 3 - 1 PTC - Per 120 m of joints, within 3 months, for same material, Per WPS for Cat. A & B joints.

UG-84 (j) – Rejection / Retesting.

UHA-51 – Impact tests for High alloy steels.

UHA-51(a) (3) – Impact values (Avg.0.38 mm & min.0.25 mm for all materials).

UHA-51(d)- Exemptions for Aus.SS base metals impact test if MDMT is -196 deg C and warmer.

UHA-51(e) – Exempton for Aus.SS welding impatc test if MDMT is -104 deg C and warmer.

PRODUCTION TEST COUPON - PTC - UG84

1. What is Production Test Coupon (PTC)?

A sample weld produced along with the production welds on main/most critically stressed weld joints of the vessel. All variables of coupon welding shall be same as that of production weld. Welded PTC will represent in all respect the properties of production weld.

2. What is the Purpose of PTC?

PTC Facilitates destructive testing of true representation of production welds for ascertaining mechanical properties.

1. **When PTC Required? - UCS 66**
   - Whenever UCS 66 curves do not exempt impact testing of Parent Metal based on
- MDMT (Minimum Design Metal Temperature) of equipment
- Thickness of Material
- Curve representing Material Specification & Grade
   OR
   - Whenever Customer Specification calls for
2. **How many PTC Required ? - UG 84**
   - 1 PTC per equipment for Categories A & B joints OR
   - As Specified In Customer Specification
   - Separate PTC required for
     - Each Welding Procedures
     - Different Welding Positions
     - Different Base Metal Thicknesses

3. **What All Testing PTC Subjected To? - UG 84**
   - Charpy V notch Impact Tests of HAZ & WELD
- 1 Set of 3 impact specimens from HAZ of all weld thicknesses and
- 1 Set of 3 impact specimens from WELD of joint thicknesses 37mm and below.
- 2 sets of 3 impact specimens each from WELD of joint thicknesses over 38 mm.
   - Other Mechanical Tests As Per Specification if Any
4. **What Test Temperature? - UG 84.1**
   - Not warmer than MDMT

5. What Test Value Acceptable? – UG 84.1

15 to 26 ft. lb / 20 to 37 ft. lb – From Fig. UG 84.1 based on minimum yield strength & Thickness of material.

OR

As specified in specification if any.

PTC Welding & Processing

1. PTC to be welded along with first long seam – Company Norm
2. PTC shall undergo all PWHT envisaged on Production welds prior to testing
3. PTC may be radiographed to eliminate possible defects from test specimens
4. PTC shall be processed / tested immediately on completion of welding
5. PTC results shall decide acceptability of production welds
6. Failure of PTC calls for rejection / repair / rework of production welds
7. PTC, therefore, demands special attention and care in handling & testing

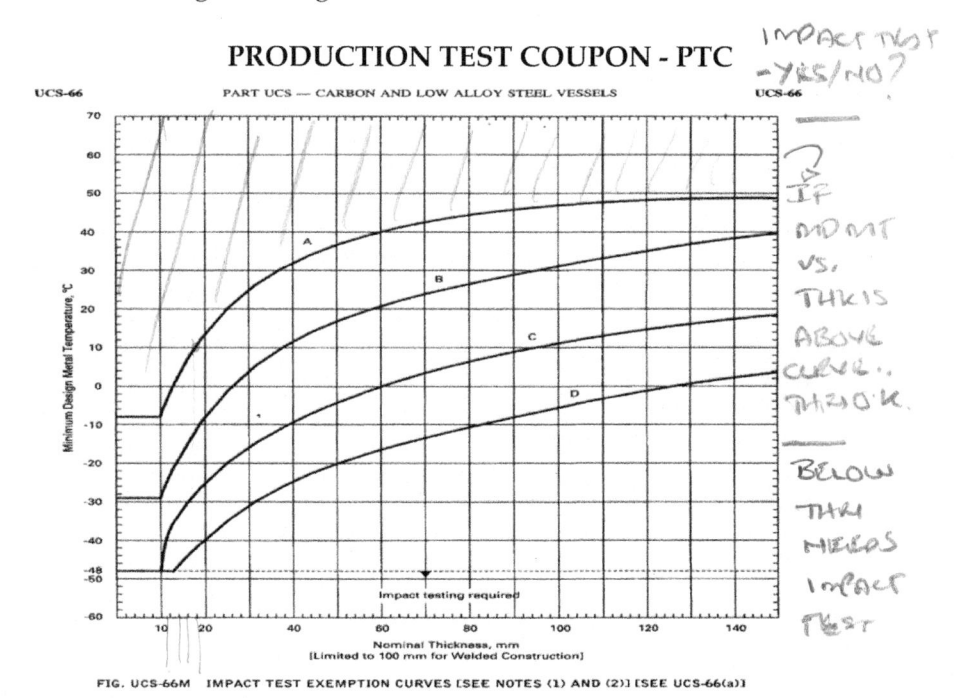

PRODUCTION TEST COUPON – PTC

[Handwritten: IMPACT TEST – YES/NO?]

[Handwritten: IF MDMT vs. THK IS ABOVE CURVE... THK IS O.K.

BELOW THRU NEEDS IMPACT TEST]

FIG. UCS-66M IMPACT TEST EXEMPTION CURVES [SEE NOTES (1) AND (2)] [SEE UCS-66(a)]

PTC
UG 84.1

2004 SECTION VIII — DIVISION 1

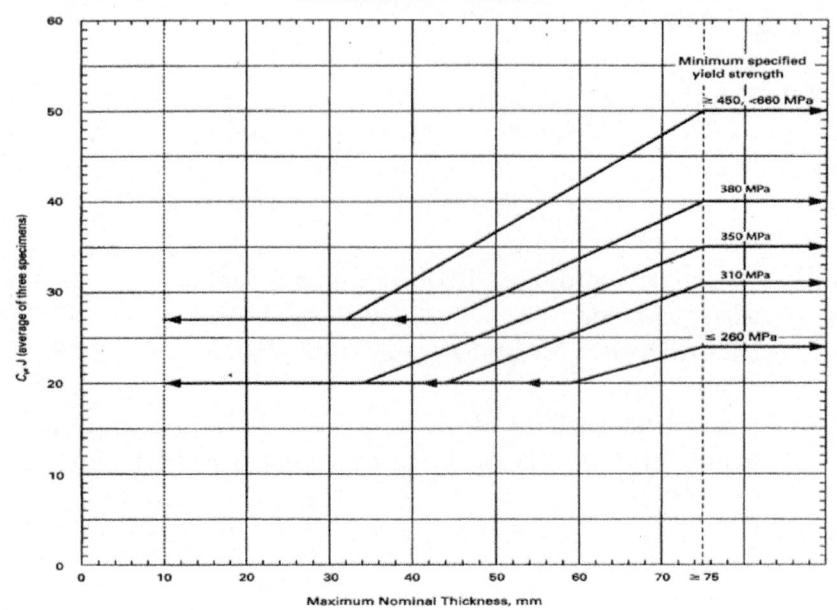

[Chart: Charpy impact energy C_v, J (average of three specimens) vs. Maximum Nominal Thickness, mm, with curves for Minimum specified yield strength: ≥450, <660 MPa; 380 MPa; 350 MPa; 310 MPa; ≤260 MPa]

GENERAL NOTES ON ASSIGNMENT OF MATERIALS TO CURVES:
(a) Curve A applies to:
   (1) all carbon and all low alloy steel plates, structural shapes, and bars not listed in Curves B, C, and D below;
   (2) SA-216 Grades WCB and WCC if normalized and tempered or water-quenched and tempered; SA-217 Grade WC6 if normalized and tempered or water-quenched and tempered.
(b) Curve B applies to:
   (1) SA-216 Grade WCA if normalized and tempered or water-quenched and tempered
      SA-216 Grades WCB and WCC for thicknesses not exceeding 2 in. (50 mm), if produced to fine grain practice and water-quenched and tempered
      SA-217 Grade WC9 if normalized and tempered
      SA-285 Grades A and B
      SA-414 Grade A
      SA-515 Grade 60
      SA-516 Grades 65 and 70 if not normalized
      SA-612 if not normalized
      SA-662 Grade B if not normalized
      SA/EN 10028-2 P295GH as-rolled;
   (2) except for cast steels, all materials of Curve A if produced to fine grain practice and normalized which are not listed in Curves C and D below;
   (3) all pipe, fittings, forgings and tubing not listed for Curves C and D below;
   (4) parts permitted under UG-11 shall be included in Curve B even when fabricated from plate that otherwise would be assigned to a different curve.
(c) Curve C applies to:
   (1) SA-182 Grades 21 and 22 if normalized and tempered
      SA-302 Grades C and D
      SA-336 F21 and F22 if normalized and tempered
      SA-387 Grades 21 and 22 if normalized and tempered
      SA-516 Grades 55 and 60 if not normalized
      SA-533 Grades B and C
      SA-662 Grade A;
   (2) all material of Curve B if produced to fine grain practice and normalized and not listed for Curve D below.
(d) Curve D applies to:
      SA-203
      SA-508 Grade 1
      SA-516 if normalized
      SA-524 Classes 1 and 2
      SA-537 Classes 1, 2, and 3
      SA-612 if normalized
      SA-662 if normalized
      SA-738 Grade A
      SA-738 Grade A with Cb and V deliberately added in accordance with the provisions of the material specification, not colder than −20°F (−29°C)
      SA-738 Grade B not colder than −20°F (−29°C)
      SA/AS 1548 Grades 7-430, 7-460, and 7-490 if normalized
      SA/EN 10028-2 P295GH if normalized [see Note (g)(3)]
      SA/EN 10028-3 P275NH

General Notes and Notes continue on next page

Location Of Impact Specimens – UG 84 (g),(h) Qualification With Supplementary Essential Variable.

Weld Joint Thickness 38 mm (1½ ") And Below

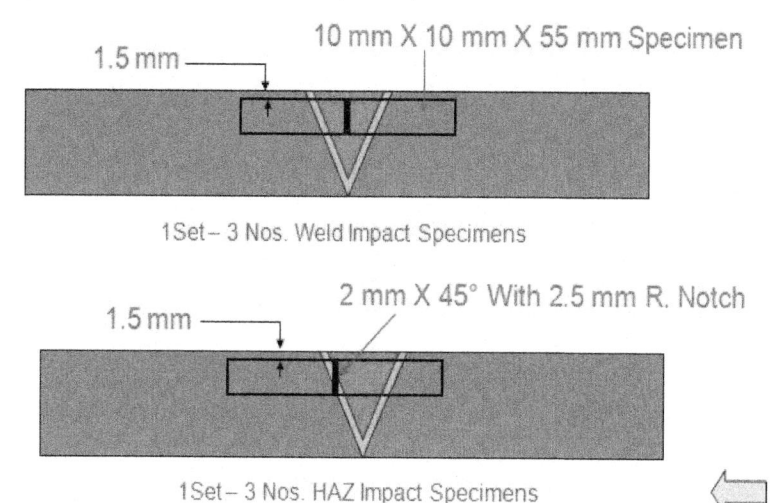

1.5 mm
10 mm X 10 mm X 55 mm Specimen

1Set – 3 Nos. Weld Impact Specimens

1.5 mm
2 mm X 45° With 2.5 mm R. Notch

1Set – 3 Nos. HAZ Impact Specimens

Weld Joint Thickness Above 38 mm (1½ ")

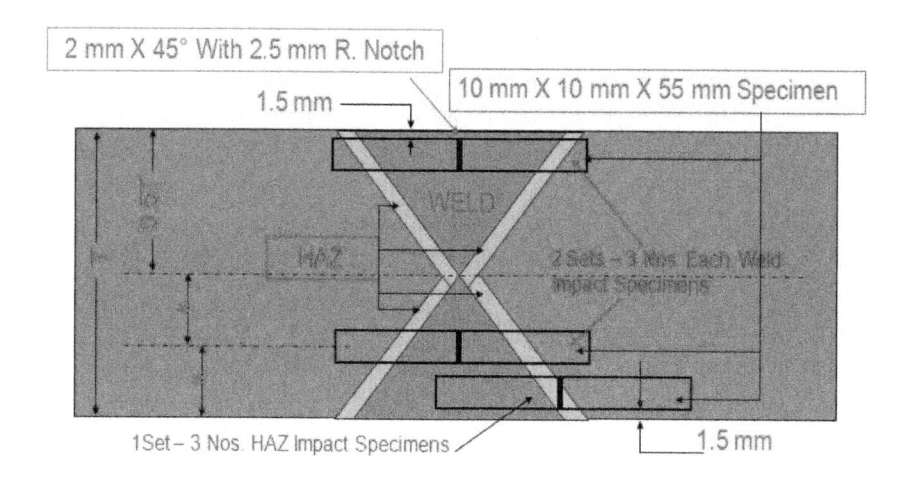

2 mm X 45° With 2.5 mm R. Notch
1.5 mm
10 mm X 10 mm X 55 mm Specimen

WELD
HAZ
2 Sets – 3 Nos. Each Weld Impact Specimens

1Set – 3 Nos. HAZ Impact Specimens
1.5 mm

MPT / PT / UT

MPT
Mandatory Appendix 6 - Methods for Magnetic Particle examination.
Relevant Indication - dimension greater than 1.5 mm.
Relevant Indication - dimension greater than 1.5 mm.
Acc.Stds. - All surfaces shall be free of: Reevant linear indication, Rounded indication greater than 5 mm, Four or more relevant rounded indications in a line seperated by 1.5 mm or less, edge to edge.

PT
Mandatory Appendix 8 - Methods for Liquid penetrant examination.
Relevant Indication - dimension greater than 1.5 mm.
Acc.Stds. - All surfaces shall be free of: Reevant linear indication, Rounded indication greater than 5 mm, Four or more relevant rounded indications in a line seperated by 1.5 mm or less, edge to edge.

UT
UW-53 - UT of Welded Joint-As per Mandatory Appendix 12
The written procedure is required.
Mandatory Appendix 12 - UT of Welds
Acc. Crit- (1) Crack, LF,IP not allowed (2) Indications exceeding ref..level amplitude and length not exceeding Max. 6 mm up to 19 mm thk. (3) max. 1/3 t for thk. 19 mm to 57 mm (4) max.19 mm for thk. Over 57 mm.

Hydro / Pneumatic Test

UG-99 - Hydrostatic Test
UG-98(a) - Max. all.work.pr. is counted at top only, for any other location static head should be considered for calculation of Max.all.work.pr.

Formula of static head by Hydraulics law (Not given in code),

For statcic head calculation of water (If the vessel is in hoz. position),the gauge pressure can be calculated as = p1 (Pascal) = (ρ g h) + p2.
where, ρ = Density of water (generally 1000 kg/m3), g = acceleration of gravity generally (9.81 m/s2), h= difference in elevation / height (Meter), p2 = Test pressure req. at top.

Hydro Test (At Top only) - 1.3 times design/ max. all. working pressure x lowest stress ratio (Stress Ratio for each material = stress value at test temp / stress value at design temp as per Sec 2D).

UG-99 (k) - vessels except for lethal ser. May be painted prior t to pressure test.

Chloride content – less than 25 ppm for ss vessels & less than 50 ppm for CS vessels.

UG-102-Test Gauges - Range -Min.1.5 times & Max. 4 times the test pressure.
UG-100 - Pneumatic test in lieu of Hydrostatic test - 1.1 times design/ working pressure x lowest stress ratio and mandatory NDT of weld as per UW-50.

OVALITY - OUT OF ROUNDNESS

Maximum Ovality Permissible On Completed Vessel = 1% Nomi. Dia - UG 80

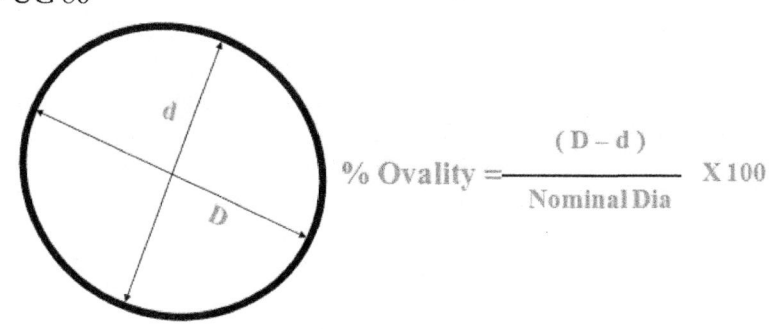

$$\% \text{ Ovality} = \frac{(D-d)}{\text{Nominal Dia}} \times 100$$

Try for 50% of Permissible at Set Up Stage – Comp. Std

OVALITY NEAR OPENING

Maximum Ovality Permissible Near Any Opening = 1% D + 2% d – UG 80

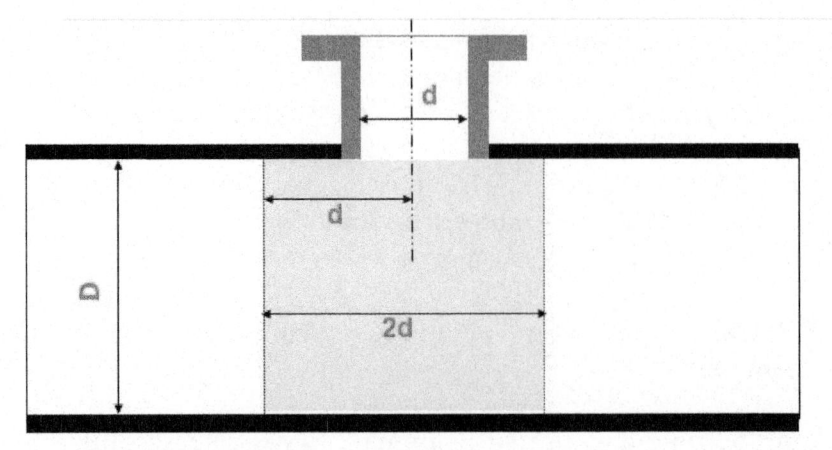

Try for 50% of Permissible at Set Up Stage – Comp. Std

PROFILE VERIFICATION OF SHELL SECTIONS UNDER EXTERNAL PRESSURE

- Ovality Check – 0.5 Ø (1 % Ø Code Requirement)
- Template Check – UG 80 (b)
    - Chord Length of Template = 2 Times the Arc Length Obtained From UG 29.2 (Based on OD, Thickness & Length)
    - Vessel Length as Given in UG 28 & 28.1
    - Gap Permissible With the Template to be Obtained from UG 80.1 (1 t to 0.2 t)
    - Hemispherical D/Es & Spherical Portion of Other D/Es, Template Check using L/D = 0.5

DERIVATION OF OUT SIDE CIRCUMFERENCE

- THE REQUIRED OUT SIDE CIRCUMFERENCE SHALL BE DERIVED BY THE FOLLOWING METHOD.

REQUIRED OUT SIDE CIRCUMFERENCE = (ID + 2T) × π
WHERE, I.D = INSIDE DIA. AS SPECIFIED IN DRAWINGS
T = ACTUAL THICKNESS MEASURED AT ENDS
PIE = VALUE OBTAINED FROM CALCULATOR

- MEASURE THE ACTUAL CIRCUMFERENCE ON THE JOB.

- ONLY CALIBRATED TAPES SHALL BE USED FOR THE PURPOSE OF MEASUREMENT.

- COMPARE THESE TWO VALUES OF CIRCUMFERENCE AND CHECK THE ACCEPTABILITY BASED ON TOLERANCE CHART.

TOLERANCES ON CIRC. OF SHELLS, D'ENDS, CONES & BELLOWS

COMPONENTS	\multicolumn{4}{c}{TOLERANCE ON CIRCUMFERENCE THICKNESS OF COMPONENTS IN MM}	REMARKS
	UP TO 8	9 TO 18	19 TO 50	ABOVE 50	
ROLLED SHELLS AT L/S SET-UP STAGE	(+) 5 (-) 2	(+) 7 (-) 4	(+) 10 (-) 7	(+) 12 (-) 9	CIRCUM. TO BE CHECKED AT 3 LOCATIONS, 2 ENDS AND AT CENTRE
PRESSED / FORMED D'END, CONE & EXPANSION BELLOW AFTER COMPLETING WELD JOINT IF ANY	(+/-) 5 (-0/+10)**	(+/-) 8 (-0/+15)**	(+/-) 10 (-0/+18)**	(+/-) 15 (-0/+25)**	IF COMPONENT IS IN SET-UP STAGE WITHOUT WELDING PER JOINT 3MM SHRINKAGE SHALL BE ADDED TO BASIC CIRCUMFERENCE.

NOTES:
- The above limits are applicable only if no others stringent tolerance is specified.
- ** this tolerance is applicable only when abutting shells has higher thickness with outside taper and inside flush.

PROFILE VERIFICATION OF SHELL SECTIONS UNDER EXTERNAL PRESSURE

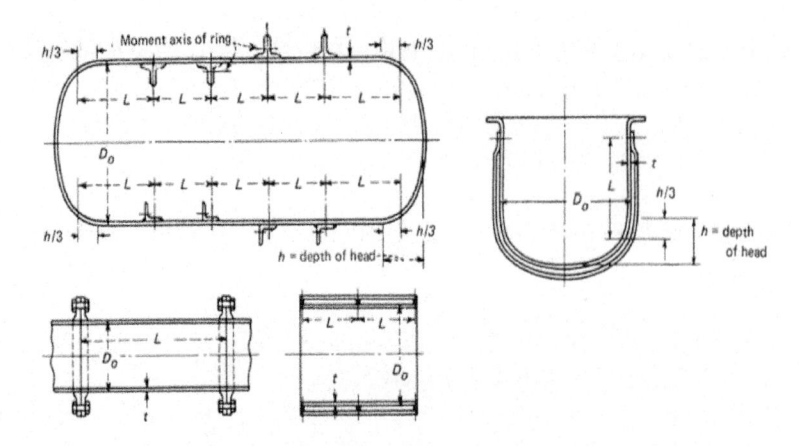

FIG. UG-28 DIAGRAMMATIC REPRESENTATION OF VARIABLES FOR DESIGN OF CYLINDRICAL VESSELS SUBJECTED TO EXTERNAL PRESSURE

PROFILE VERIFICATION OF SHELL SECTIONS UNDER EXTERNAL PRESSURE

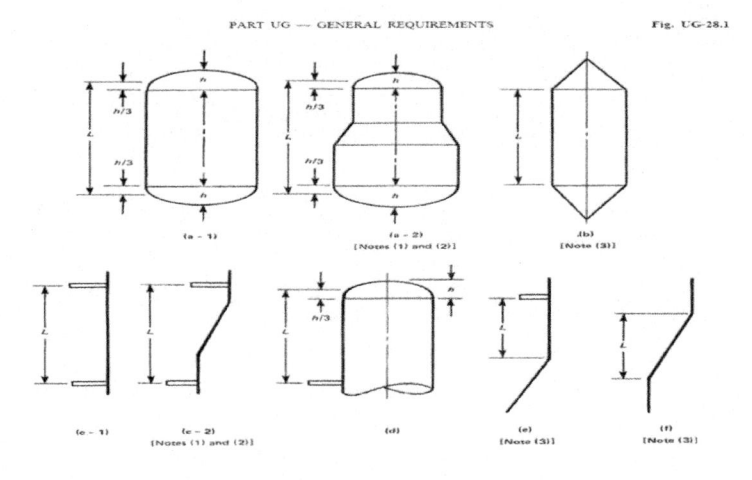

NOTES:
(1) When the cone-to-cylinder or the knuckle-to-cylinder junction is not a line of support, the nominal thickness of the cone, knuckle, or toriconical section shall not be less than the minimum required thickness of the adjacent cylindrical shell.
(2) Calculations shall be made using the diameter and corresponding thickness of each section with dimension L as shown.
(3) When the cone-to-cylinder or the knuckle-to-cylinder junction is a line of support, the moment of inertia shall be provided in accordance with 1-8.

FIG. UG-28.1 DIAGRAMMATIC REPRESENTATION OF LINES OF SUPPORT FOR DESIGN OF CYLINDRICAL VESSELS SUBJECTED TO EXTERNAL PRESSURE

PROFILE VERIFICATION OF SHELL SECTIONS UNDER EXTERNAL PRESSURE

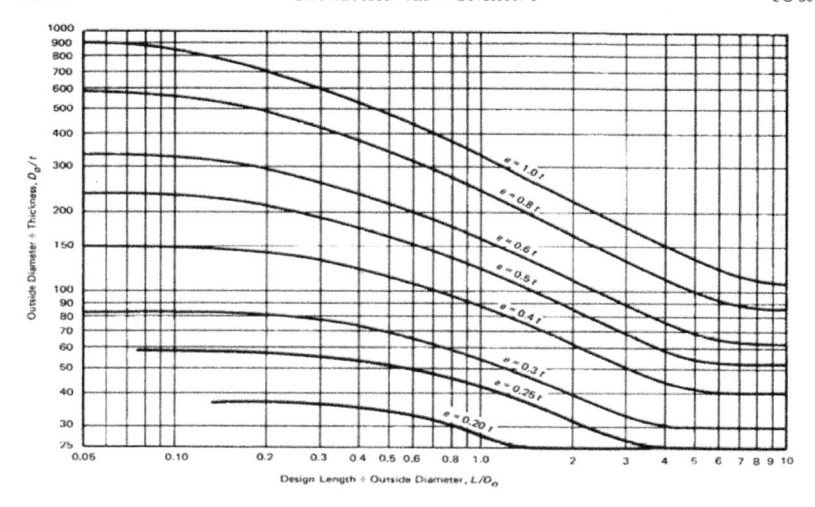

FIG. UG-80.1 MAXIMUM PERMISSIBLE DEVIATION FROM A CIRCULAR FORM e FOR VESSELS UNDER EXTERNAL PRESSURE

PROFILE VERIFICATION OF SHELL SECTIONS UNDER EXTERNAL PRESSURE

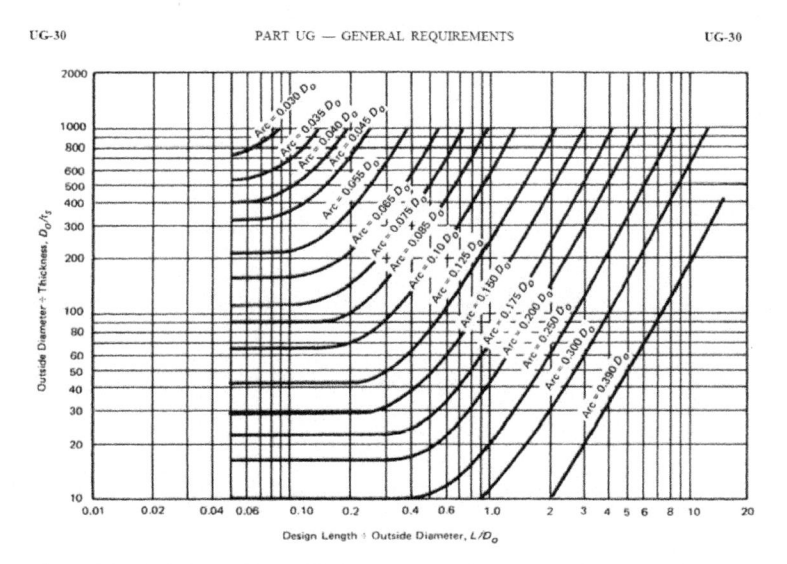

FIG. UG-29.2 MAXIMUM ARC OF SHELL LEFT UNSUPPORTED BECAUSE OF GAP IN STIFFENING RING OF CYLINDRICAL SHELL UNDER EXTERNAL PRESSURE

INTERNAL PRESSURE DESIGN

Thickness of shells under Internal Pressure (UG-27)

- All dimensions in code are in corroded condition
- Provision for any loading
- **Symbols:**
  - t = min. required thickness
  - P = internal design pressure
  - R = inside radius of shell
  - S = max. allowable stress value (Sec II D)
  - E = joint efficiency (UW-12)

Cylindrical Shells (UG-27)

1. **Circumferential Stress (Longitudinal Joints).**
   $t \le 0.5R$, $\quad P \le 0.385SE$

$$t = \frac{PR}{SE - 0.6P} \quad \text{or} \quad P = \frac{SEt}{R + 0.6t}$$

2. **Longitudinal Stress (Circumferential Joints).**
   $t \le 0.5R$, $\quad P \le 1.25SE$

$$t = \frac{PR}{2SE + 0.4P} \quad \text{or} \quad P = \frac{2SEt}{R - 0.4t}$$

Thickness Calculation based on Outside diameter (Appendix 1-1)

1. **Circumferential Stress (Longitudinal Joints).**

$$t = \frac{PR_0}{SE + 0.4P} \quad \text{or} \quad P = \frac{SEt}{R_0 - 0.4t}$$

Where R_0 = outside radius of shell

2. **Longitudinal Stress (Circumferential Joints).**

$$t = \frac{PR_0}{2SE + 1.4P} \quad \text{or} \quad P = \frac{2SEt}{R_0 - 1.4t}$$

Spherical Shells (UG-27)
Stresses in biaxial direction are same.

$$t \le 0.356R, \quad P \le 0.665SE$$

$$t = \frac{PR}{2SE - 0.2P} \quad \text{or} \quad P = \frac{2SEt}{R + 0.2t}$$

Based on Outside Diameter (Appendix 1-1)

$$t = \frac{PR_0}{2SE + 0.8P} \quad \text{or} \quad P = \frac{2SEt}{R_0 - 0.8t}$$

Conical Shells Sections (UG-32)

α = one half of the included (apex) angle of the cone ($\alpha < 30°$)
D = inside diameter at point under consideration.
D_o = outside diameter at point under consideration (Appendix 1-4).

1. Circumferential Stress (Longitudinal Joints)

$$t = \frac{PD}{2\cos\alpha(SE - 0.6P)} \quad \text{or} \quad P = \frac{2SEt\cos\alpha}{D + 1.2t\cos\alpha}$$

$$t = \frac{PD_0}{2\cos\alpha(SE + 0.4P)} \quad \text{or} \quad P = \frac{2SEt\cos\alpha}{D_0 - 0.8t\cos\alpha}$$

2. Longitudinal Stress (Circumferential Joints).

$$t = \frac{PD}{4\cos\alpha(SE + 0.4P)} \quad \text{or} \quad P = \frac{4SEt\cos\alpha}{D - 0.8t\cos\alpha}$$

$$t = \frac{PD_0}{4\cos\alpha(SE + 1.4P)} \quad \text{or} \quad P = \frac{4SEt\cos\alpha}{D_0 - 2.8t\cos\alpha}$$

Formed Heads

- UG-32 –Formed Heads
    - -Ellipsoidal
    - -Torispherical
    - -Hemispherical
    - -Conical
    - -Toriconical

Formed Heads

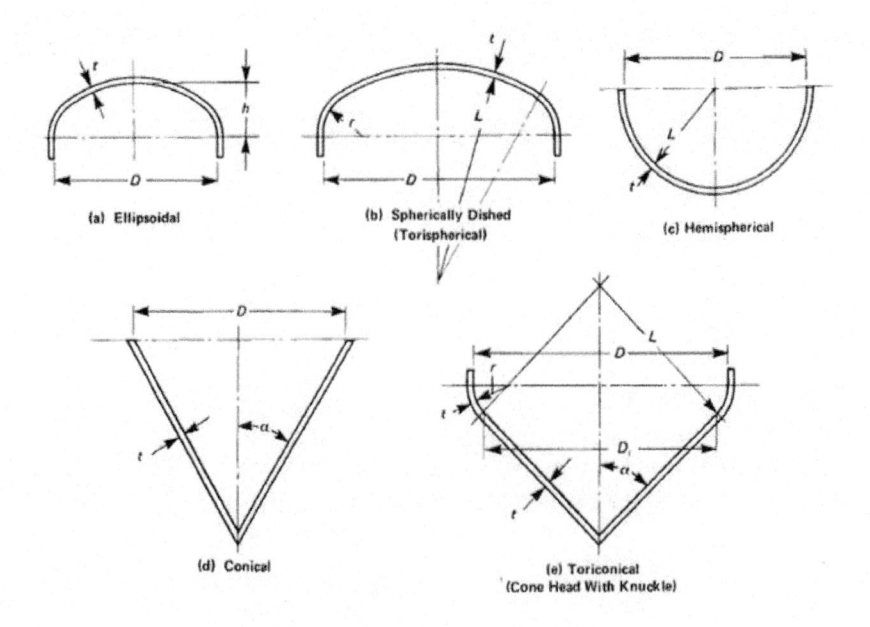

FIG. 1-4 PRINCIPAL DIMENSIONS OF TYPICAL HEADS

(a) Ellipsoidal
(b) Spherically Dished (Torispherical)
(c) Hemispherical
(d) Conical
(e) Toriconical (Cone Head With Knuckle)

Ellipsoidal Heads (UG-32) (Appendix 1-4)

t = minimum required thickness of head after forming
P = internal design pressure
D = inside diameter of the head skirt, or inside length of the major axis pf an ellipsoidal head
S = max. allowable stress value in tension (Sec II D)
E = lowest efficiency of any joint in the head
D_0 = outside diameter of the head skirt, or outside length of the major axis of an ellipsoidal head
h = the inside depth of the ellipsoidal head measured from the tangent line
K = a factor in formulas for ellipsoidal heads depending on the head proportion $D/2h$.

Ellipsoidal Heads (UG-32) (Appendix 1-4)
Table 1-4.1

Values of factor K
(Use nearest Values of D/2h; Interpolation Unnecessary)

D/2h	3.0	2.9	2.8	2.7	2.6	2.5	2.4	2.3	2.2	2.2	2.0
K	1.83	1.73	1.64	1.55	1.46	1.37	1.29	1.21	1.14	1.07	1.00
D/2h	1.9	1.8	1.7	1.6	1.5	1.4	1.3	1.2	1.1	1.0	...
K	0.93	0.87	0.81	0.76	0.71	0.66	0.61	0.57	0.53	0.50	...

Ellipsoidal Heads (UG-32) (Appendix 1-4) (Continued…)

- $t/L \geq 0.002$
- Inside Depth = ¼ Inside diameter

$$t = \frac{PD}{2SE - 0.2P} \quad \text{or} \quad P = \frac{2SEt}{D + 0.2t}$$

- 2:1 ellipsoidal Head-
    Approximation: Knuckle Radius = 0.17D
    Spherical Radius = 0.90D

Ellipsoidal Heads (UG-32) (Appendix 1-4) (Continued…)

$t/L \geq 0.002$

$$t = \frac{PDK}{2SE - 0.2P} \quad \text{or} \quad P = \frac{2SEt}{KD + 0.2t}$$

$$t = \frac{PD_0 K}{2SE + 2P(K - 0.1)} \quad \text{or} \quad P = \frac{2SEt}{KD_0 - 2t(K - 0.1)}$$

Where $K = \frac{1}{6}\left[2 + \left(\frac{D}{2h}\right)^2\right]$

Torispherical Heads (UG-32) (Appendix 1-4)

t = min. required thickness of head after forming
P = internal design pressure
D = inside diameter of the head skirt,
S = max. allowable stress value in tension (Sec II D)
E = lowest efficiency of any joint in the head
r = Inside Knuckle Radius
L = inside spherical or crown radius
D_o = outside diameter of the head skirt,
L_0 = outside spherical or Crown radius
M = a factor in formulas for Torispherical heads depending on the head proportion L/r

Table 1-4.2
Values of factor M
(Use nearest Values of L/r; Interpolation Unnecessary)

L/r	1.0	1.25	1.50	1.75	2.00	2.25	2.50	2.75	3.00	3.25	3.50
M	1.00	1.03	1.06	1.08	1.10	1.13	1.15	1.17	1.18	1.20	1.22
L/r	4.0	4.5	5.0	5.5	6.0	6.5	7.0	7.5	8.0	8.5	9.0
M	1.25	1.28	1.31	1.34	1.36	1.39	1.41	1.44	1.46	1.48	1.50
L/r	9.5	10.00	10.5	11.0	11.5	12.0	13.0	14.0	15.0	16.0	$16\tfrac{2}{3}$[1]
M	1.52	1.54	1.56	1.58	1.60	1.62	1.65	1.69	1.72	1.75	1.77

Notes -
1) Maximum ratio allowed by UG-32(j) when L equals the outside diameter of the skirt of the head.

Torispherical Heads (UG-32) (Appendix 1-4) (Continued...)
- $t/L \geq 0.002$
- Knuckle radius = 6% of inside crown radius
- Inside crown radius = outside diameter of skirt

$$t = \frac{0.885PL}{SE - 0.1P} \quad \text{or} \quad P = \frac{SEt}{0.885L + 0.1t}$$

- Where minimum tensile strength > 70,000 psi

$$S = 20000 \times \frac{\text{Allowable Stress Value at Temprature}}{\text{Allowable Stress Value at Room Temprature}}$$

Torispherical Heads (UG-32) (Appendix 1-4) (Continued...)

$$t = \frac{PLM}{2SE - 0.2P} \quad \text{or} \quad P = \frac{2SEt}{LM + 0.2t}$$

$$t = \frac{PL_0M}{2SE + P(M - 0.2)} \quad \text{or} \quad P = \frac{2SEt}{ML_0 - t(M - 0.2)}$$

Where $M = \frac{1}{4}\left(3 + \sqrt{\frac{L}{r}}\right)$

Hemispherical Heads (UG-32) (Appendix 1-4)
- $t \leq 0.356L, \quad P \leq 0.665SE$

$$t = \frac{PL}{2SE - 0.2P} \quad \text{or} \quad P = \frac{2SEt}{L + 0.2t}$$

- L = inside spherical radius

Sample Problem
Design of Vertical Cylindrical Pressure Vessel for Internal Pressure

Design Data
- ❖ Design pressure   : 12 Kg/cm²g
- ❖ Design temperature : 250° C
- ❖ Radiography   : Spot
- ❖ Corrosion Allowance : 3 mm
- ❖ Material   : SA 515 Gr. 70
- ❖ Inside Diameter   : 2500 mm
- ❖ Tan to Tan length   : 10000 mm
- ❖ Liquid head/ density : 4000mm from bottom tan line/ 1200 Kg/cu.m

Calculate the Following :
- ❖ Shell Thickness
- ❖ Thickness of 2:1 ellipsoidal dished head
- ❖ Thickness of Torispherical head with crown radius 2500 mm and Knuckle radius 300 mm
- ❖ Thickness of Hemispherical head

- ❖ Shell Thickness as per UG-27[c] (Solution)

- ❖ Design pressure   P=12 Kg/cm²g = 0.12 kg/mm²
- ❖ Corroded Inside radius   R=1250mm
- ❖ Corrosion Allowance   C= 3 mm
- ❖ Joint Efficiency   E=0.85
- ❖ Allowable stress   S=

S for SA 515 Gr.70 at 482°F (250°C) is 20000 psi (As per Section II D)
Now, 1 pound = 0.45359237 Kg.

$$S = \frac{20 \times 10^3 \times 0.4536}{(25.4)^2} \approx 14.062 \, Kg/mm^2$$

Total Pressure at Bottom Tan line P = Internal Pressure + Liquid Pressure
Liquid Pressure = ρ*g*h
        = 1200 x 9.81 x 4

$$= 47088 \frac{kg}{m^2} \times \frac{m}{\sec^2}$$

= 47088 N/m²
= 47088/9.81x1000000 Kg/mm²
= 0.0048 kg/mm²
so, Total Design Pressure P= 0.12+0.0048 = 0.1248 kg/mm²

- ❖ Required Thickness for Circumferential Stress
  P= 0.1248 ≤ 0.385SE=4.6 kg/mm² (So condition Satisfied)

$$t = \frac{PR}{SE - 0.6P} + C = \frac{0.1248 \times 1253}{(14.062 \times 0.85 - 0.6 \times 0.1248)} + 3$$

$$\approx 13.13 + 3$$
$$\approx 16.13 mm$$

- ❖ Required Thickness for Longitudinal Stress
  P= 0.1248 ≤ 1.25SE=14.94 kg/mm² (So condition Satisfied)

$$t = \frac{PR}{2SE + 0.4P} + C = \frac{0.1248 \times 1253}{(2 \times 14.062 \times 0.85 + 0.4 \times 0.1248)} + 3$$

$$\approx 6.512 + 3$$
$$\approx 9.512 mm$$

- **2:1 Ellipsoidal head as per UG-32 (Solution)**
- Design pressure   P=0.12 Kg/mm²
- Corroded Inside radius   R=1253mm
- Corrosion Allowance   C= 3 mm
- Joint Efficiency   E=0.85
- Allowable stress   S= 14.062 kg/mm² at 482°F (Sec-IID)
- **Inside Depth** $h = \dfrac{D}{4} + 3 = \dfrac{2500}{4} + 3 = 628 mm$
- **Factor K** $K = \dfrac{1}{6}\left[2 + \left(\dfrac{D}{2h}\right)^2\right] = \dfrac{1}{6}\left[2 + \left(\dfrac{2506}{2 \times 628}\right)^2\right]$

  $\approx 1 mm$

- **Required Thickness at top**

$$t = \dfrac{PDK}{2SE - 0.2P} + C = \dfrac{0.12 \times 2506 \times 1}{2 \times 14.062 \times 0.85 - 0.2 \times 0.12} + 3$$

$$\approx 15.59 mm$$

- **Torispherical head as per UG-32 (Solution)**
- Design pressure   P=0.12 Kg/mm²
- Corroded Inside Diameter D=2506mm
- Corrosion Allowance   C= 3 mm
- Joint Efficiency   E=0.85
- Allowable stress   S= 14.062 kg/mm² at 482°F (Sec-IID)
- Corroded Crown Radius   L≈ D (2500 mm)
- Corroded Knuckle Radius   r>6% of Inside Diameter
  (consider r=300 mm)
- **Factor M** $M = \dfrac{1}{4}\left(3 + \sqrt{\dfrac{L}{r}}\right) = \dfrac{1}{4}\left(3 + \sqrt{\dfrac{2500}{300}}\right)$

  $\approx 1.47 mm$

- Required thickness at top

$$t = \dfrac{PLM}{2SE - 0.2P} + C = \dfrac{0.12 \times 2500 \times 1.47}{2 \times 14.062 \times 0.85 - 0.2 \times 0.12} + 3$$

$$\approx 21.47 mm$$

Sample Problem (Continued…)

- ❖ Hemispherical head as per UG-32 (Solution)
- ❖ Design pressure at bottom P=12+[5.25*1.2/10]=0.1263 Kg/mm²
- ❖ Corroded Inside Diameter   D=2506mm
- ❖ Corrosion Allowance   C= 3 mm
- ❖ Joint Efficiency   E=0.85
- ❖ Allowable stress   S= 14.062 kg/mm² at 482°F (Sec-IID)
- ❖ Corroded Crown Radius   L=1253 (D/2)
- ❖ Required thickness

$$t = \frac{PL}{2SE - 0.2P} + C = \frac{0.1263 \times 1253}{2 \times 14.062 \times 0.85 - 0.2 \times 0.1263} + 3$$

Design for External Pressure and compressive stresses

- Compressive forces caused by dead weight, wind, earthquake, internal vacuum
- Can cause elastic instability (buckling)
- Vessel must have adequate stiffness
  – Extra thickness
  – Circumferential stiffening rings
- ASME procedures for cylindrical shells, heads, conical sections. Function of:
    – Material
    – Diameter
    – Unstiffened length
    – Temperature
    – Thickness

TABLE 1A (CONT'D)
SECTION I; SECTION III, CLASS 2 AND 3;* AND SECTION VIII, DIVISION 1
MAXIMUM ALLOWABLE STRESS VALUES S FOR FERROUS MATERIALS
(*See Maximum Temperature Limits for Restrictions on Class)

	Line No.	Nominal Composition	Product Form	Spec No.	Type/Grade	Alloy Desig./ UNS No.	Class/ Cond./ Temper	Size/ Thickness, in.	P-No.	Group No.
	1	C-Si	Castings	SA-352	LCB	J03003	...	...	1	1
A99	2	C-Si	Plate	SA-515	65	K02800	...	...	1	1
A99	3	...	...	...	...	...	...	...	...	...
	4	C-Mn-Si	Plate	SA-516	65	K02403	...	...	1	1
A99	5	C-Si	Wld. pipe	SA-671	CB65	K02800	...	...	1	1
A99	6	C-Mn-Si	Wld. pipe	SA-671	CC65	K02403	...	...	1	1
A99	7	C-Si	Wld. pipe	SA-672	B65	K02800	...	...	1	1
A99	8	C-Mn-Si	Wld. pipe	SA-672	C65	K02403	...	...	1	1
	9	C-Mn	Sheet	SA-414	E	K02704	...	...	1	1
	10	C-Mn-Si	Plate	SA-662	B	K02203	...	...	1	1
A99	11	C-Mn-Si	Plate	SA-537	...	K12437	1	$2\frac{1}{2} < t \leq 4$	1	2
A99	12	C-Mn-Si	Wld. pipe	SA-691	CMSH-70	K12437	...	$2\frac{1}{2} < t \leq 4$	1	2
A99	13	C-Si	Forgings	SA-226	2	K03506	...	...	1	2
	14	C-Mn	Plate	SA-455	...	K03300	...	$0.58 < t \leq \frac{3}{4}$	1	2
	15	C	Bar	SA-675	70	...	...	...	1	2
	16	C-Si	Forgings	SA-105	...	K03504	...	...	1	2
	17	C-Si	Forgings	SA-181	...	K03502	70	...	1	2
	18	C-Si	Castings	SA-216	WCB	J03002	...	...	1	2
A99	19	C-Si	Forgings	SA-266	2	K03506	...	...	1	2
	20	C-Mn-Si	Forgings	SA-266	4	K03017	...	...	1	2
	21	C-Mn-Si	Forgings	SA-350	LF2	K03011	...	...	1	2
	22	C-Si	Forgings	SA-508	1	K13502	...	...	1	2
	23	C-Mn-Si	Forgings	SA-508	1A	K13502	...	...	1	2
	24	C-Si	Forgings	SA-541	1	K03506	...	...	1	2
	25	C-Mn-Si	Forgings	SA-541	1A	K03506	...	...	1	2
A99	26	C-Si	Cast pipe	SA-660	WCB	J03003	...	...	1	2
	27	C-Mn-Si	Forgings	SA-765	II	K03047	...	...	1	2
	28	C-Si	Plate	SA-515	70	K03101	...	...	1	2
	29	C-Mn-Si	Plate	SA-516	70	K02700	...	...	1	2
A99	30	C-Si	Wld. pipe	SA-671	CB70	K03101	...	...	1	2
A99	31	C-Mn-Si	Wld. pipe	SA-671	CC70	K02700	...	...	1	2
A99	32	C-Si	Wld. pipe	SA-672	B70	K03101	...	...	1	2
A99	33	C-Mn-Si	Wld. pipe	SA-672	C70	K02700	...	...	1	2
	34	C-Si	Smls. pipe	SA-106	C	K03501	...	...	1	2
A99	35	C-Mn-Si	Wld. tube	SA-178	D	...	...	...	1	2
A99	36	C-Mn-Si	Wld. tube	SA-178	D	...	...	...	1	2
	37	C-Mn-Si	Wld. tube	SA-178	D	...	...	...	1	2
	38	C-Mn-Si	Smls. tube	SA-210	C	K03501	...	...	1	2
	39	C-Mn-Si	Castings	SA-216	WCC	J02503	...	...	1	2

TABLE 1A (CONT'D)
SECTION I; SECTION III, CLASS 2 AND 3;* AND SECTION VIII, DIVISION 1
MAXIMUM ALLOWABLE STRESS VALUES S FOR FERROUS MATERIALS
(*See Maximum Temperature Limits for Restrictions on Class)

Line No.	Min. Tensile Strength, ksi	Min. Yield Strength, ksi	I	III	VIII-1	External Pressure Chart No.	Notes	
1	65	35	NP	700	650	CS-2	G1, G17	
2	65	35	1000	700	1000	CS-2	G10, S1, T2	A99
3	...	...	...	...	...	...	...	A99
4	65	35	850	700	1000	CS-2	G10, S1, T2	
5	65	35	NP	700	NP	CS-2	S6, W10, W12	A99
6	65	35	NP	700	NP	CS-2	S6, W10, W12	A99
7	65	35	NP	700	NP	CS-2	S6, W10, W12	A99
8	65	35	NP	700	NP	CS-2	S6, W10, W12	A99
9	65	38	NP	NP	900	CS-2	G10, G35, T1	
10	65	40	NP	NP	700	CS-3	T1	
11	65	45	NP	700	650	CS-4	T1	A99
12	65	45	NP	700	NP	CS-4	G26, T1, W10, W12	A99
13	70	35	NP	NP	1000	CS-2	G10, T2	A99
14	70	35	NP	400 (Cl. 3 only)	650	CS-2	...	
15	70	35	850	650 (Cl. 3 only)	1000	CS-2	G10, G15, G18, G22, G35, S1, T2	
16	70	36	1000	700	1000	CS-2	G10, G18, G35, S1, T2	
17	70	36	1000	700	1000	CS-2	G10, G18, G35, S1, T2	
18	70	36	1000	700	1000	CS-2	G1, G10, G17, G18, S1, T2	
19	70	36	1000	700	NP	CS-2	G10, G18, S1, T2	A99
20	70	36	NP	NP	1000	CS-2	G10, T2	
21	70	36	NP	700	1000	CS-2	G10, T2	
22	70	36	NP	700	1000	CS-2	G10, T2	
23	70	36	NP	700	1000	CS-2	G10, T2	
24	70	36	NP	700	1000	CS-2	G10, T2	
25	70	36	NP	700	1000	CS-2	G10, T2	
26	70	36	1000	700	NP	CS-2	G1, G10, G17, G18, S1, T2	A99
27	70	36	NP	NP	650	CS-2	...	
28	70	38	1000	700	1000	CS-2	G10, S1, T2	
29	70	38	850	700	1000	CS-2	G10, S1, T2	
30	70	38	NP	700	NP	CS-2	S5, W10, W12	A99
31	70	38	NP	700	NP	CS-2	S6, W10, W12	A99
32	70	38	NP	700	NP	CS-2	S5, W10, W12	A99
33	70	38	NP	700	NP	CS-2	S6, W10, W12	A99
34	70	40	1000	700	1000	CS-3	G10, S1, T1	
35	70	40	1000	NP	NP	...	G10, S1, T1, W14	A99
36	70	40	1000	NP	NP	...	G4, G10, S1, T4	A99
37	70	40	1000	NP	NP	...	G3, G10, S1, T2	
38	70	40	1000	NP	1000	CS-3	G10, S1, T1	
39	70	40	1000	700	1000	CS-3	G1, G10, G17, G18, S1, T1	

TABLE 1A (CONT'D)
SECTION I; SECTION III, CLASS 2 AND 3;* AND SECTION VIII, DIVISION 1
MAXIMUM ALLOWABLE STRESS VALUES S FOR FERROUS MATERIALS
(*See Maximum Temperature Limits for Restrictions on Class)

Maximum Allowable Stress, ksi (Multiply by 1000 to Obtain psi), for Metal Temperature, °F, Not Exceeding

	Line No.	−20 to 100	150	200	250	300	400	500	600	650	700	750	800	850	900
	1	18.6	18.6	18.6	...	18.6	18.6	18.6	17.9	17.3	16.7	...	...	...	...
A99	2	18.6	18.6	18.6	...	18.6	18.6	18.6	17.9	17.3	16.7	13.9	11.4	8.7	5.9
A99	3	...	...	...	...	...	...	...	...	...	...	...	...	...	...
	4	18.6	18.6	18.6	...	18.6	18.6	18.6	17.9	17.3	16.7	13.9	11.4	8.7	5.9
A99	5	18.6	...	18.6	...	18.6	18.6	18.6	17.9	17.3	16.7	...	...	...	...
A99	6	18.6	...	18.6	...	18.6	18.6	18.6	17.9	17.3	16.7	...	...	...	...
A99	7	18.6	...	18.6	...	18.6	18.6	18.6	17.9	17.3	16.7	...	...	...	...
A99	8	18.6	...	18.6	...	18.6	18.6	18.6	17.9	17.3	16.7	...	...	...	...
	9	18.6	18.6	18.6	...	18.6	18.6	18.6	18.6	18.6	16.9	13.9	11.4	8.7	5.9
	10	18.6	18.6	18.6	...	18.6	18.6	18.6	18.6	18.6	16.9	...	...	...	...
A99	11	18.6	...	18.6	...	18.6	18.6	18.6	18.6	18.6	16.9	...	...	...	...
A99	12	18.6	...	18.6	...	18.6	18.6	18.6	18.6	18.6	16.9	...	...	...	...
A99	13	20.0	20.0	20.0	...	20.0	19.9	19.0	17.9	17.3	16.7	14.8	12.0	9.3	6.7
	14	20.0	20.0	20.0	...	20.0	19.9	19.0	17.9	17.3	...	...	...	...	...
	15	20.0	20.0	20.0	...	20.0	19.9	19.0	17.9	17.3	16.7	14.8	12.0	9.3	6.7
	16	20.0	20.0	20.0	...	20.0	20.0	19.6	18.4	17.8	17.2	14.8	12.0	9.3	6.7
	17	20.0	20.0	20.0	...	20.0	20.0	19.6	18.4	17.8	17.2	14.8	12.0	9.3	6.7
	18	20.0	20.0	20.0	...	20.0	20.0	19.6	18.4	17.8	17.2	14.8	12.0	9.3	6.7
A99	19	20.0	...	20.0	...	20.0	20.0	19.6	18.4	17.8	17.2	14.8	12.0	9.3	6.7
	20	20.0	20.0	20.0	...	20.0	20.0	19.6	18.4	17.8	17.2	14.8	12.0	9.3	6.7
	21	20.0	20.0	20.0	...	20.0	20.0	19.6	18.4	17.8	17.2	14.8	12.0	9.3	6.7
	22	20.0	20.0	20.0	...	20.0	20.0	19.6	18.4	17.8	17.2	14.8	12.0	9.3	6.7
	23	20.0	20.0	20.0	...	20.0	20.0	19.6	18.4	17.8	17.2	14.8	12.0	9.3	6.7
	24	20.0	20.0	20.0	...	20.0	20.0	19.6	18.4	17.8	17.2	14.8	12.0	9.3	6.7
	25	20.0	20.0	20.0	...	20.0	20.0	19.6	18.4	17.8	17.2	14.8	12.0	9.3	6.7
A99	26	20.0	...	20.0	...	20.0	20.0	19.6	18.4	17.8	17.2	14.8	12.0	9.3	6.7
	27	20.0	20.0	20.0	...	20.0	20.0	19.6	18.4	17.8	...	...	...	...	...
	28	20.0	20.0	20.0	...	20.0	20.0	20.0	19.4	18.8	18.1	14.8	12.0	9.3	6.7
	29	20.0	20.0	20.0	...	20.0	20.0	20.0	19.4	18.8	18.1	14.8	12.0	9.3	6.7
A99	30	20.0	...	20.0	...	20.0	20.0	20.0	19.4	18.8	18.1	...	...	...	...
A99	31	20.0	...	20.0	...	20.0	20.0	20.0	19.4	18.8	18.1	...	...	...	...
A99	32	20.0	...	20.0	...	20.0	20.0	20.0	19.4	18.8	18.1	...	...	...	...
A99	33	20.0	...	20.0	...	20.0	20.0	20.0	19.4	18.8	18.1	...	...	...	...
	34	20.0	...	20.0	...	20.0	20.0	20.0	20.0	19.8	18.3	14.8	12.0	9.3	6.7
A99	35	20.0	...	20.0	...	20.0	20.0	20.0	20.0	19.8	18.3	14.8	12.0	9.3	6.7
A99	36	20.0	...	20.0	...	20.0	20.0	20.0	20.0	19.8	18.3	14.8	12.0	9.3	5.7
	37	17.0	...	17.0	...	17.0	17.0	17.0	17.0	16.8	15.5	12.6	10.2	7.9	5.7
	38	20.0	...	20.0	...	20.0	20.0	20.0	20.0	19.8	18.3	14.8	12.0	9.3	6.7
	39	20.0	20.0	20.0	...	20.0	20.0	20.0	20.0	19.8	18.3	14.8	12.0	9.3	6.7

Note: if any changes in Code and specification kindly take update from official website.
Kindly give ratting star and comment your experience after buy this book.

Thank You

ABOUT THE AUTHOR
QA./QC Engineer
Technical, industrial, Engineering, Quality Control Knowledgeable *and jobs releted ebooks and* handbook *Writer.*

Printed in Great Britain
by Amazon